Modeling Nature

Science and Its Conceptual Foundations

DAVID L. HULL, EDITOR

Modeling Nature

Episodes in the History of Population Ecology

Sharon E. Kingsland

The University of Chicago Press
CHICAGO AND LONDON

Sharon E. Kingsland is an assistant professor
of the history of science at the Johns Hopkins University.

The University of Chicago Press, Chicago 60637
The University of Chicago Press, Ltd., London

Printed in the United States of America
94 93 92 91 90 89 88 87 86 85   5 4 3 2 1

Library of Congress Cataloging in Publication Data

Kingsland, Sharon E.
Modeling nature.

(Science and its conceptual foundations)
Bibliography: p.
Includes index.
1. Population biology—History. 2. Popı
biology—Mathematical models—History. J
II. Series.
QH352.K56 1985 574.5′248
ISBN 0-226-43726-4

To Paul

Contents

Portrait Photographs

Acknowledgments

Near the end of *Middlemarch*, George Eliot observed that "there was no need to praise anybody for writing a book, since it was always done by somebody else." Of all those who helped me to write this book, I am most indebted to G. Evelyn Hutchinson, who provided extensive commentary on the whole manuscript, and to Mary P. Winsor, whose influence as a teacher has been too pervasive to be calculated.

Several biologists took time to talk to me about their work or to read parts of the manuscript in various stages. I thank Robert Bakker, Jacques Berger, Robert E. Cook, Charles Elton, G. F. Gause, Michael Hassell, Henry Horn, Jeremy B. C. Jackson, John Lawton, Robert M. May, Robert P. McIntosh, Thomas Park, R. Chris Plowright, Henry Regier, Oscar W. Richards, O. W. Richards, H. N. Southern, Gary Sprules, and George C. Varley.

David Hull and William C. Wimsatt provided criticism of some of the philosophical portions; Malcolm Kottler commented on chapter seven; Paul M. Romney gave advice on many points of style. Frank Egerton and the late Luisa Volterra D'Ancona supplied information about A. J. Nicholson and Vito Volterra, respectively. Gail Schley typed part of the manuscript. To all these people I am most grateful.

I thank the Princeton University Library (Special Collections) for permission to quote from the A. J. Lotka Papers; and the American Philosophical Society Library for permission to quote from the Raymond Pearl Papers. This book incorporates previously published material from the *Quarterly Review of Biology*, vol. 57, 1982, pages 29–52; and *Human Biology*, vol. 56, 1984, pages 1–18. I thank the Stony Brook Foundation and the Wayne State University Press for permission to use this material.

September 1984

Introduction

Ecology is the study of patterns in nature, of how those patterns came to be, how they change in space and time, why some are more fragile than others. Population ecology is concerned with how populations interact with the environment and how these interactions give rise to the larger patterns of communities and ecosystems. The environment is more than just sun, air, earth, and water: it includes other organisms which may help or hinder the survival of a species. Population ecology is also the study of how those organisms interact, through food webs, in competition, and in cooperation. The population itself is often defined arbitrarily to fit the convenience of the investigator. It may be a group of oak trees in a forest, a swarm of insects in a field, a cluster of flies in a bottle, or a wholly abstract entity existing only on a mathematician's page. This book is about the places where the imagined world of the mathematician and the real world of the biologist intersect. The debates generated by this interaction involve the problem of knowing how to identify patterns.

Today population ecology overlaps with several related biological fields, including economic biology, biogeography, sociobiology, epidemiology, and population genetics. The overlap between population ecology and population genetics began relatively recently, in the 1960s, although both fields trace their origins to the 1920s. But back then these two aspects of population biology were distinct, partly from the sheer difficulty of combining them and partly because they had distinct historical origins. Population genetics concerned the changes in gene frequencies in populations: for simplicity, the environment was assumed to be invariable and ecological interactions unimportant. Population ecology concerned the effects of changing environmental conditions on a population: to make the problem manageable, the populations were assumed not to vary genetically. Where mathematical models were used, even the physical environment was assumed to be unchanging, and the only variables were the numbers of organisms in the populations. Population ecology grew up alongside population genetics and confronted many of the same problems as its companion, problems involving the use of mathematics and the difficulties of integrating theoretical, experimental, and field approaches.

But it was a separate area of study, the other side of the coin, the analysis of the day-to-day interactions that make up the struggle for existence.

Population ecology, as a distinct branch of ecology, has had its own struggles for existence. In the late nineteenth century, when ecology was itself an embryonic discipline, ecologists were led to study populations in an attempt to unravel the complexities of the food web. By the 1920s, when animal ecologists were actively moving to separate themselves from the dominating influence of plant ecologists, the study of populations took on a new importance. People recognized that attention to the individual organism and its interaction with the environment was artificially restrictive. It omitted too many kinds of interactions with the living world. Attempts to understand the structure of communities, on the other hand, were leading to unwieldy masses of observations with few organizing principles to guide research. The middle level, the population, became an attractive focus of attention, all the more attractive because understanding population fluctuations was crucial to the sound management of agricultural resources.

While ecologists began to study populations, both in the field and in the laboratory, others began to explore the possibilities of a mathematical representation of population growth, predation, and competition. These others were not ecologists: their ideas and methods came from the physical sciences and from the human sciences, especially from demography. Their work was received with some interest, and a great deal of skepticism. Mathematical models were drastically oversimplified: what did a set of equations have to do with the moiling crowd of the real world? But they did give people ideas, and a few enterprising ecologists set out to test them experimentally and to apply them to field research. The mathematical studies began to seep into the ecological literature. By the 1940s they had gained scarcely more than a foothold, but it was a firm one.

By the mid-forties population ecologists formed an active community within ecology. A few centers of population research, which had led a precarious existence during the thirties, found themselves in a secure position after the war. Population ecology's permanence was declared. But this new part of ecology was still an odd hybrid, compounded of several distinct disciplines, and roughly divided into three branches corresponding to field, laboratory, and theoretical investigation. These branches formed an uneasy alliance, with the theoretical side still on the fringes. Population ecology had developed in a curious manner. Those responsible for its establishment—for the journals, the texts, the symposia, the institutions devoted to population research—were committed to field and laboratory research, to long-term, laborious studies with a Baconian cast of fact gathering about them. But part of the content of population ecology, its theoretical content, had entered from without, constructed by good mathematicians who often had little sense of the biological. There the theory sat, waiting to be assimilated in bits, creating lasting tensions between ecologists leaning to one side or the other.

Following the war, population ecologists paused to take stock of the new directions in which their field was being propelled. Upon reflection, it was immediately clear that there would be no simple answers to the questions of population regulation and community structure that had been puzzled over for so long. The search for simple "laws," always a tempting exercise for ecologists in envy of the physical sciences, would have to give way to the development of more sophisticated techniques of analysis. These would require not only more biological information, but more mathematics and eventually the aid of computers. It was also clear that ecologists were arguing at cross purposes because too much of their science had been erected on a foundation of ill-defined concepts. Vagueness is not intrinsically harmful: many advances are made by the introduction of a metaphor which at first seems to bring a problem to order, but upon more careful scrutiny turns out to be maddeningly hard to pin down. It may be more harmful to try to define one's terms too early in the development of a field. But ecologists were too splintered to tolerate continued vagueness in the terms they used every day. Not only were they uneasy about the theoretical part of their science, they were not at all agreed as to the status of the facts themselves. And so in the 1950s they stopped to review, to criticize, to define, and to argue some more. They worked to pick out what was most useful and to integrate the theoretical findings a little better with their observations. In trying to remove ambiguities, they argued incessantly over what words meant and whether a given idea signified something real or illusory. Attention was on the real world and on exploring its complexities. At the same time, in the wake of the evolutionary synthesis of the previous decade, some population ecologists began to deal more directly with problems in evolution. Ecology was becoming associated with neo-Darwinism.

Ecology, always a fragmented discipline, was by now entering an expansive phase which would produce by the late 1960s a science largely reconstituted as ecological engineering. Recognition of worldwide environmental problems spurred the formation of the International Biological Program in 1960. The United States National Committee, appointed by the National Academy of Sciences in 1965, coordinated until 1974 a program of ecological research of unprecedented scope. Emphasis was on large cooperative projects which sought to analyze whole ecosystems. Studies of energy flow, biogeochemistry, and radiation studies (many funded by the Atomic Energy Commission) took on a prominent role. The use of new instruments and techniques was accompanied by new rigor in the use of mathematics. Not only were ecologists becoming versed in quantitative and statistical biology, but in collaboration with mathematicians they were turning also to more esoteric branches of mathematics which were growing quickly during and after the war years. These included systems theory, information theory, mathematical programming, and game theory.

In population dynamics more work was being done to develop realistic and precise models which could be used for long-term prediction in economic biology. The areas of fish population dynamics and entomology were in the forefront in the use of statistical methods, mathematical modeling, and life-table analysis. Solutions to complex problems appeared to be mainly a question of increasing the level of sophistication of models and the methods of data collection. The possibility that a predictive ecology was in the offing was increased by the promise of the new computer technology. By the mid-1960s ecologists were becoming interested in computer simulations of ecological interactions traced over the course of several generations. Within a few years computers would become indispensable in population dynamics. The goals of management, prediction, and control of nature seemed finally to be within reach.

But while one part of ecology, interested in solving practical problems and encouraged by the new computer technology, aimed at getting more precise models, there was another theoretical line developing which was disarmingly simple. Its principal author was Robert H. MacArthur, though much of the theory was developed through collaboration. His models were more general: they gave no numerical predictions, but distinguished between alternatives only in a qualitative sense. MacArthur dared to suggest that answers to rather large questions could be found by reducing a problem to its barest essentials and choosing between a limited number of alternative explanations. His ideas stimulated more research and invited controversy. Mathematical ecology gained strength under MacArthur's impact and became allied with evolutionary biology and genetics. But it continued to provoke controversy through the 1960s and 1970s and well into the present day. The question was not whether mathematics was useful: it was obvious that some mathematics was necessary for these difficult inquiries. Rather the question was, how far might one stray from the real world before running the risk of losing sight of it entirely? Where was the right balance between the mathematical view of the world and the more focused view of the ecologist studying nature in a particular time and place? Rhetoric abounded, but beneath it lay serious concerns about the future of ecology as a science, not of one uniform style, but of many.

This conflict is itself an ancient one in science and has to do with the purpose of scientific inquiry. On the one hand, knowledge may be sought for purely practical reasons, to predict and control some part of nature for society's benefit. On the other hand, knowledge may serve more abstract ends for the contemplative soul. Uncovering new relationships is aesthetically satisfying in that it brings order to a chaotic world. In ecology these goals are intertwined and not easily separable. Different emphases do sometimes dictate different strategies and different attitudes toward the abstract reasoning of mathematics. But there is no final end toward which this all tends, no fixed answer that serves as a marker of progress: what is sought is a better understanding of the connectedness of nature's parts. The history of ecology is a history of changing criteria for imposing order on nature and resisting the alternative that all is really chaotic and contingent. Ecology is interesting not just for the answers it comes up with, for these are often temporary, but for the way the methods of imposing upon nature reflect changing times, changing moods.

In addition, in ecology as distinct from most other areas of the biological and physical sciences, many of the questions asked have traditionally been viewed historically. What we observe is the culmination of a unique series of events which can be pieced together only to the extent that there remain traces in the historical record, or to the extent that past processes can be inferred from presently acting ones. Mathematical models of the kind that I shall mostly discuss are not historical in this sense. They are forward looking rather than backward looking. Their aim is not to reconstruct history. Rather they use the present situation to create a plausible, if oversimplified, scenario and then try to predict some outcome. Failure of the prediction to be borne out by observation may indicate that the assumptions of the model are inappropriate. The general goal of the mathematical method is not at odds with the investigative historical method. Both are intended to uncover more information about the world. But the mathematical way of thinking, that is, the use of models to construct plausible scenarios, while commonplace in the physical sciences, is in its ahistorical character opposed to the way of thinking familiar to most ecologists. The difficulty of trying to reconcile these two ways of thinking has been the source of much controversy.

Apart from the philosophical questions concerning the purpose of science and the different ways we can choose to make inquiries of nature, the debates which recur in population ecology also have an economic, or, if you like, an ecological aspect. Framed in ecological terms the issue is, quite simply, how many schools of ecology can peacefully coexist? This type of problem may be resolved historically by diversification and specialization within a discipline, and in ecology such specialization has occurred. But it may also be resolved by the elimination of one or more schools, thereby lessening the pressures on limited financial resources. Some of the conflicts described in this book stemmed precisely from this feeling that one's survival was being threatened by an opposing school.

The following chapters present not a comprehensive history of population ecology, but a series of episodes which illustrate variations on these recurring themes. A great deal has been left out. My emphasis is on the theoretical side of population ecology, especially animal ecology. I have ignored the history of plant ecology, as well as the growth of ecosystem ecology during this period. I have also not fleshed out the recent history of economic biology where mathematical modeling has become very important. Nor have I given full weight to the huge amount of experimental and field research on populations which, though important for the history of ecology, did not enter directly into the few episodes I have chosen. For the same reason I have not discussed those other areas of biology such as ethology which are crucial for the understanding of the modern field, but which had little impact upon population ecology in the period I am covering.

Some of these episodes do not involve ecologists at all, but are interesting in their own right and were important later on in helping to set the foundations for population ecology. Many ecologists might regard the bulk of this book as the "prehistory" of population ecology. To some extent it is that, but it is nevertheless interesting to see how, although the context is not ecology strictly speaking, the patterns and processes of these earlier times do shed light on the modern discipline.

The first chapter provides a sketch of a few of the leading ideas of ecology, by way of introduction to the problems of population analysis. From there I jump to an anomalous figure in the history of science, Alfred J. Lotka, whose contributions to ecology were made almost despite himself. Lotka was part of that turn-of-the-century generation of Americans who, reacting partly to the views of Herbert Spencer, embraced a vision of man as an active part of an interconnected cosmos. The idea of nature as a unity including humans, influenced by German philosophy, was reflected in science by the attempt to discover the laws governing that unity, and by an interest in such questions as the origins of consciousness and the individual's changing role in a technological society. It was also reflected in a desire to break down the artificial barriers between the social sciences, which were just emerging, and the established natural sciences. Lotka was fully part of these early twentieth-century trends, and his holistic visions expressed many of the themes of his age. Though not explicitly ecological, his writings pulled at the fringes of the new ecological science. At a time when the biological and social sciences were growing specialized, ecologists were working to break down these barriers and to forge a synthetic discipline which would include the study of the human species as part of nature.

Man as an animal was also the theme of Raymond Pearl, who may be said to have "discovered" Lotka. His work is the subject of the third and fourth chapters. Though Pearl was not an ecologist either, his attempts to create a human biology, blending the statistical and demographic methods of the human sciences with the experimental methods of the biological sciences, were allied with the trends in ecology in the 1920s toward viewing human and animal societies from a common perspective. Pearl's work also rested on the edges of ecology, especially as ecology turned more toward the study of populations. Of particular interest is the extraordinary controversy surrounding Pearl's logistic curve, which he touted as a law of population growth in the 1920s. Population biology has always been full of polemic, but ecologists in our modern, relatively restrained age might be surprised at the strength with which biologists in Pearl's day hurled barbs at each other.

From Pearl I move to a broader discussion of mathematical modeling in chapter five. This brings a return to Lotka and to his physicist colleague Vito Volterra, who by virtue of his high reputation in mathematical physics succeeded in impinging upon ecology where Lotka had failed. From there the focus shifts briefly to Australia to consider the somewhat oddly reasoned theories of Alexander J. Nicholson, who used an elaborate mechanical analogy to develop a theory of population regulation and competition.

In chapter six I review ecologists' responses to these outpourings of the twenties and thirties. Although some were enthusiastic, most were cautious in appraising this mathematical corpus. One critic whose vehemence was rather a surprise was William R. Thompson. He had himself pioneered the use of mathematical models in his youthful days, and he had inspired Lotka to develop his predator-prey model. Thompson's change of attitude is an interesting study of the tensions created when the mathematician's models must be turned to practical use and are found to conflict with the results of long and laborious field study.

In chapter seven I discuss how mathematical theory influenced the development of certain key ideas in ecology and evolutionary biology. The chapter begins with a discussion of G. F. Gause's experimental studies of Volterra's models. This young Russian ecologist was also influenced by Raymond Pearl, who helped him to publish his first book in America. Gause's studies were meant to integrate ecological and evolutionary processes, though he was unsuccessful in creating a synthesis of these two levels. But his ideas, embodied in the principle that two species cannot occupy the same niche, influenced David Lack when he later came to study the ecological specialization of the Galápagos finches. Lack's ornithological studies, coming in the wake of the modern evolutionary synthesis, helped to make ecologists more aware of the need to address evolutionary questions by looking at life-history strategies. Meanwhile in America, G. Evelyn Hutchinson was using Gause's work to elaborate his own ideas about competition and the niche.

In chapter eight I move from a discussion of Hutchinson's point of view in population ecology to that of his stellar student, Robert H. MacArthur. MacArthur's way of thinking about ecological problems both reinvigorated ecology and set off new debates on the place of theoretical reasoning in this descriptive science. His work was both brilliant and controversial: it is too early yet to assess its full impact. I present here a sketch of parts of his research, with some assessments from his scientific colleagues. The conclusion summarizes some of the main themes of the book, themes having to do with philosophical conflicts and the patterns of social interaction within science.

One way to see the gradual development of population ecology is not in terms of a conflict between field, laboratory, and theory, or between the different goals of pure and applied biology, but as a conflict between historical and ahistorical thinking. Mathematical ecology was intended to address historical issues: short-term ecological changes and long-term evolutionary changes. It was the theory of the struggle for existence. But the very act of imposing mathematics (or any model) on nature often involved a rejection of history in favor of a harmonious, unifying concept. Robert MacArthur's methods best illustrated this deliberate choice of problems which would minimize the importance of history—would minimize, that is, explanations which focused on unique, historical events—in order to generate new hypotheses. This was what he meant by his frequent insistence that ecology is the search for patterns of repetition. He therefore put especial emphasis on populations in equilibrium, rather than populations which fluctuated in an apparently erratic fashion. But he was not the first to develop an equilibrium view of nature: such a view has been common wherever ideas taken from physics have been transferred to a biological context. When an equilibrium approach is applied to ecology and evolution, the attempt to minimize historical explanations can seem dangerously misleading. Therefore in response to the mathematical analysis of equilibrium cases, biologists have from time to time stressed the need to recapture the insights of historical awareness. The dialectic between historical and ahistorical perceptions continues to unfold and to cause controversy, but along with it comes a better understanding of the complexity of nature and of the role of theoretical models in biology.

1

Prologue: The Entangled Bank

Darwin was wonderfully struck by how the presence of the most complex forms of creation could be explained by referring to the common processes continually in action and open to our view. Closing *On the Origin of Species*, he wrote:

> It is interesting to contemplate an entangled bank, clothed with many plants of many kinds, with birds singing on the bushes, with various insects flitting about, and with worms crawling through the damp earth, and to reflect that these elaborately constructed forms, so different from each other, and dependent on each other in so complex a manner, have all been produced by laws acting around us.[1]

The most important of these laws was the tendency of populations to increase in a geometric ratio, an increase which would lead inevitably to a struggle for existence. A few calculations helped to emphasize the hidden power of nature. Linnaeus, Darwin tells us, figured that an annual plant producing two seeds, with each of its offspring doing the same, would in twenty years produce a million plants. Darwin writes that he himself took some pains to calculate that a single pair of elephants would have at least fifteen million descendants after five centuries. In the manuscript version of the "Struggle for Existence" chapter, many more such figures are compiled. Darwin clearly hoped to impress his readers with the irrefutable logic of fat, solid numbers.[2]

Such calculations, however whimsical an image of the future they gave, were necessary because in reality these rates of increase are seldom achieved, and never for any great length of time. Apart from periodic fluctuations and occasional disturbances, the face of nature looks fairly uniform from one year to the next. The calculations of nature's potential were needed to stress the inevitability of the struggle for existence. As Darwin wrote, "Nothing is easier than to admit in words the truth of the universal struggle for life, or more difficult—at least I have found it so—than constantly to bear this conclusion in mind."[3] The struggle for existence, in its large metaphorical sense, included not only the obvious cases of competition between near equals, but also any dependence of one organism on another or on external conditions. The struggle encompassed all the ecological relations which might affect an individual's survival or the survival of its offspring.

A different metaphor conveys how Darwin saw the struggle in action. I quote here from the longer and more forceful version of Darwin's manuscript, rather than the lines that appeared in *Origin of Species*, in order to give the full flavor of Darwin's visual imagination:

> Nature may be compared to a surface covered with ten-thousand sharp wedges, many of the same shape & many of different shapes representing different species, all packed closely together & all driven in by incessant blows: the blows being far severer at one time than at another; sometimes a wedge of one form & sometimes another being struck; the one driven deeply in forcing out others; with the jar and shock often transmitted very far to other wedges in many lines of direction: beneath the surface we may suppose that there lies a hard layer, fluctuating in its level, & which may represent the minimum amount of food required by each living being, & which layer will be impenetrable by the sharpest wedge.[4]

The comparison of species to wedges, with the powers of increase represented by the incessant blows, constitutes a puzzling metaphor. But the picture powerfully conveys Darwin's sense of nature as full, tightly packed, with each species sunk more or less firmly into its place in the surface of the world. As each species drives itself in, it nudges the ones closest, sending repercussions outward and perhaps forcing another species out. While these waves are being felt, opportunities may arise for new species to fill the places in the economy of nature created by these interactions.

The idea that a struggle for existence is a prerequisite for the origin of species by natural selection is tied to the idea of nature as finely balanced. Species are prevented from realizing their powers of increase by the presence of innumerable checks, which exact a heavy toll in death. Darwin saw these checks as arising mostly from the presence of other species, rather than from the physical conditions alone. In the metaphor, it is the combined presence of other wedges—other species—rather than the resistance of the substrate, which is most crucial in determining whether a wedge can be driven inward. The influence of other organisms is subtle and surprising: cattle determine the presence of scotch fir on the open heath; the cats in a district, by controlling the mice, which control the bees, can be held responsible for the numbers of flowers in the area. There is a balance in nature, but it is a delicate balance, liable to be upset by the smallest grain.

The delicacy of this balance explains why, although we cannot foresee how natural selection will act, we can appreciate that it can indeed result in the creation of new species. All that is required is some slight change to shift the ecological balance. The shift causes new places in nature to open up, and if a favorable variation should arise that allows a population to take advantage of the new place, it will quickly seize it. In the process, through competition, it drives out the forms lacking the beneficial variation. Speciation is accompanied by extinction; the result is a continual readjustment of nature to new ecological balances. The important idea is that speciation occurs because all beings are bound up, not with the habitat alone, but in complex relations with other beings. The greatest amount of speciation will be in places where there is greatest struggle; on large, continental land masses, as opposed to small, isolated areas. On the continents, Darwin felt, the populations would be large, the ecology complex, the struggle intense, and more new places would be formed for new species to fill.

This understanding of how natural selection occurs led to a change in Darwin's concept of adaptation. Species are not adapted only to physical conditions. Rather, they are adapted to their co-inhabitants, to those on which they feed, from which they must escape, and with which they compete.[5] Adaptation is therefore a relative concept, one that can be understood only by knowing the full range of ecological relations, and one that may change rapidly and subtly as those relations change.

It is crucial for Darwin's argument that the reader keep these ecological complexities, and this idea of relative adaptation, in mind. Otherwise the working of natural selection cannot be appreciated. But for Darwin the details of these ecological relations were almost wholly unknown; at best they could be only dimly perceived. All Darwin could argue was that we should not allow ignorance to form the basis of any skepticism about natural selection. If we do not know what mechanisms are involved, we should at least have faith that nature is sufficiently complicated to achieve the results of speciation. If nothing else, we have to be convinced of our profound ignorance of nature, "a conviction as necessary, as it seems to be difficult to acquire."[6]

As a first step toward making the struggle for existence less obscure by subjecting it to scientific study, one of the early German Darwinians, Ernst Haeckel, defined a new science of *ökologie* in 1866. The word means the study of households; Haeckel intended it to stand for the science of the relations of the organism to the environment, including the study of all the conditions of existence, both organic and inorganic.[7] Haeckel was himself not an ecologist, and having given this new science a name, he did nothing more to advance its growth. But a few German botanists at the time did begin to explore what Haeckel had called ecology, although they called themselves biologists and not ecologists. These botanists were interested in applying physiology to the study of plant adaptation, thereby turning what had been a morphological study into a dynamic experimental study.[8]

The new scientific botany was closely allied with physiology, but it meant something more than "physiology proper," which was the study of the chemical and physical processes of the plant. Botany in its larger sense included the physiology of adaptation, focusing on the relations between organism and environment, and on the mutual adaptations and relations between organisms. In the 1880s these ideas began to influence American botanists as well: it was they who decided to distinguish the wider sense of physiological botany by calling it "ecology" in the 1890s.

The move to direct observation of nature, whether in morphology or physiology, was reinforced by trends in science education. In England, T. H. Huxley was championing the study of biology, the "Science of Individual Life," by the same methods of observation and experiment used in the physical sciences. Biology was to be learned not from books but from close observation, dissection, and comparison. In America, Louis Agassiz urged the same: his Museum of Comparative Zoology at Harvard, founded in 1859 amid great energy and excitement, would set new standards in zoological education. Agassiz's students felt they were part of a great historic undertaking, the pursuit of natural history "in its true sense, earnestly and scientifically," with "unlimited license in the use of specimens."[9]

These ideas for a more scientific approach to natural history and to science education were eagerly embraced elsewhere in America following the Civil War. One young man who responded enthusiastically was Stephen Alfred Forbes, who in 1872 was just starting as curator of the Museum of the Illinois State Natural History Society.[10] When he was seventeen years old, Forbes's formal education had been interrupted by army service during the war. Afterward he tried a medical course, but quit partly from poverty and partly from distaste for surgical work performed without anesthetic. His scientific interests, shaped by his agricultural background, brought him eventually back to the study of natural history, a field where it was just possible to scratch out a living.

Forbes was caught up in the sweep of ideas stimulated by Darwin, Huxley, and Agassiz, which reached Illinois in the mid-1870s.[11] In 1875 he helped to set up a summer school of natural history in Illinois, following the lead of Louis Agassiz's school on Penikese Island, Massachusetts, which had opened in 1873. The second session of Forbes's school was not run until 1878; by this time he had also founded the Illinois State Laboratory of Natural History, which turned into the State Natural History Survey in 1917. In 1882 he was appointed to serve as the fourth state entomologist of Illinois. This position had been created in 1867, when only two other states, New York and Missouri, had state entomologists. Two years later, in 1884, Forbes joined the faculty of the University of Illinois as head of the Department of Zoology and Entomology.

As a biologist working for the state, Forbes was naturally concerned with problems of importance to the agricultural sciences. Indeed, he was careful to stress the practical benefits to be gained by a more systematic, quantitative survey of the biological world, for the purely intellectual rewards of such work were hardly equal to the laborious nature of the work itself. He conjured up an image of a typical student's state of mind as he approached the tasks at hand, making it clear that any such student ought to have a well-developed sense of public duty to provide emotional sustenance in this line of work: "His material is in the worst possible condition for study; and the personal result of his labor is a continual discouragement to him. That whatever individual impulse should have been turned in this direction should have been exhausted long before definite or conclusive results were reached, was, therefore, inevitable. . . . In short, this is emphatically one of those questions which, if studied exhaustively at all, must be studied chiefly in the public interest."[12]

The question in this case was the food of birds, or more precisely, the determination of the relation of birds to agriculture through a quantitative study of their food habits. Forbes understood that before any claims could be made about the value of any species, injurious or beneficial, to society, it was essential to know how the different classes of animals were related through the food web. His studies of the food relations of birds, insects, and fish were distinguished attempts to give agriculture a scientific basis.

In addition to the economic value of these ecological surveys, Forbes's writings were distinguished also for their broad, theoretical outlook. To gather only facts was not his intention: Forbes felt it was important to cast those facts in the light of the latest theories; theories which would help organize biologists' ideas of how nature maintained its balance. The leading ideas he drew upon to guide his perceptions came from two sources, each of equal importance in shaping the growth of the budding science of ecology. These were Charles Darwin and Herbert Spencer.

From Darwin, Forbes accepted the argument that species and higher groups evolved by natural selection. He justified his work partly on the grounds that it would shed light on problems of distribution, the causes of variation, the origin and extinction of species, and the nature of adaptation. By systematic investigation, the causes and effects of the struggle for existence would be revealed. Knowing the complications of the struggle for existence raised more problems than it solved, however, for it still needed to be explained how this struggle produced a well-regulated world. Nature *did* seem to be in balance; behind the first chaotic impressions of a world in continual strife, order and lawfulness were to be discerned. Forbes wanted to know how that order was achieved, and he found his answer in the works of Herbert Spencer.

Spencer's vast writings imposed on the world a consistent vision based on the principle that all processes could be understood in terms of the balance of forces. The foundations for this worldview were set out in *First Principles*, published in 1862 and later reissued with Spencer's customary amendments and additions.[13] In *First Principles*, Spencer referred all phenomena to the persistence of a fundamental quantity, which he called "force." The use of the word "force" was rather vague: the term applied equally to the energy emanating from the sun and to such quantities as vital force, social force, and economic force, which were not intended to be metaphors, but actual forces.[14] These were the manifestations of the "Absolute Force," which we can never truly know but are only dimly aware of as the correlate of those forces that we do know.[15] These lesser forces, in their diverse forms, are known to us through our knowledge of matter and motion. It is the absolute, unknowable force that persists, or is conserved, but we can observe the results of the persistence of force through the transformations that the other forces continually undergo. The problem therefore was to discover a formula expressing the combined consequences of these forces. Because the forces could be analyzed in terms of matter and motion, the problem reduced to a search for a law of the continuous redistribution of matter and motion. In Spencer's words, the question to be answered was: "What dynamic principle, true of the metamorphosis as a whole and in its details, expresses these ever-changing relations?"[16]

The answer to the problem lay in two basic processes that Spencer called evolution and dissolution. Evolution was the integration of matter and concomitant dissipation of motion, while dissolution was the absorption of motion and disintegration of matter.[17] These processes were in constant antagonism, now one gaining the upper hand, now the other, so that "every detail of the history [of every aggregate] is definable as a part of either the one change or the other."[18] On the whole, these two processes never perfectly balanced each other, with the result that rhythmic oscillations were a universal feature of natural phenomena.

Spencer knew that his narrow definition of evolution as the integration of matter and dissipation of motion was apt to be confusing, because the term "evolution" as generally understood implied much more: it implied an unfolding or expanding. Accordingly he distinguished between simple evolution, which corresponded to the above definition, and compound evolution, which involved other changes in addition to the basic integration of matter and dissipation of motion. Spencer perceived these secondary changes to be the gradual movement from an indefinite, homogeneous state to a more specialized, heterogeneous state. His final formula for evolution was the following:

> Evolution is an integration of matter and concomitant dissipation of motion; during which the matter passes from an indefinite, incoherent

> homogeneity to a definite, coherent heterogeneity; and during which the retained motion undergoes a parallel transformation.[19]

This definition was intended to apply not only to organic evolution, but to all levels of transformation, including the evolution of the solar system, of the earth's geological structure, and of human civilization.

Having developed his basic worldview in *First Principles*, Spencer proceeded to its application. He turned to biological problems in *The Principles of Biology*, published in 1866. The final section of this work, entitled "Laws of Multiplication," bore particular relevance to ecological questions.[20] Spencer began by reasoning that the preservation of races implied a stable equilibrium between destructive forces (all those contributing to mortality) and preservative forces (those contributing to individual life or to the production of offspring). As with the other opposing forces, the antagonistic forces of population increase and decrease were necessarily balanced in a way that produced rhythmical movements about a stable equilibrium value. In addition, he noted that there had to be a further adjustment between the individual's ability to maintain its life and its ability to reproduce: high mortality was associated with high fertility, while low mortality accompanied low fertility. This adjustment produced a second general antagonism between what Spencer called individuation and genesis; that is, the processes by which individual life was maintained were opposed to the processes leading to reproduction.

Considering the second antagonism between individual maintenance and reproduction, Spencer analyzed it, using explicitly economic terms, as a balance between costs, where costs were determined by the sum total of quantities expended in the various physiological and behavioral functions. Large animals, such as the elephant, which require a great deal for individual maintenance, produce only a few offspring in a lifetime. Smaller species, with a lower grade of evolutionary development, have a correspondingly greater fertility. Indirectly, this antagonism was molded by natural selection and the survival of the fittest, for any species unable to maintain an economical balance would die out. But in day-to-day terms the economics itself controlled the ability of natural selection to produce changes: natural selection could not, all else equal, produce an increase both of fertility and of individual maintenance cost at the same time.[21]

The evolution of forms therefore implied some sort of economy, or as Spencer put it, "The vital capital invested in the alteration must bring a more than equivalent return."[22] An evolutionary change must result in the more economical performance of an action, as might be achieved, for example, through the division of labor: this would create a greater surplus of vital capital, part of which would go to the maintenance of the individual and part into reproduction.

These principles applied equally to humans. Although the pressure of population growth was the original cause of human evolution, it was also the case that the development of human skill and intelligence necessitated a decline in fertility: the higher evolution of society was at the expense of reproduction. Gradually a stable equilibrium would be reached between man's inner nature and his outer relations with society and with the environment. The attainment of this equilibrium would coincide with the equilibrium of population through the balance of births and deaths. Just as in nature, all changes worked toward a harmonious state where antagonistic forces were balanced and which was ultimately referable to the economic relations between individual and reproductive costs.[23]

These conclusions were derived by deductive argument. Although Spencer did try to strengthen them with examples from nature, the evidence was suggestive rather than conclusive. He acknowledged the difficulties of verifying his assorted deductions, but felt that there was enough evidence to support the inference that these antagonisms between forces were real. The appeal of his point of view was that it provided a way to understand natural selection in terms of the laws governing the individual, laws based on economic principles. Forbes found this explanatory power attractive: no less attractive was the fact that the explanation it offered was conducive to his vision of a harmonious, progressive world.

Forbes accepted Spencer's principle that the species was maintained at the cost of the individual.[24] From this basic law of nature, he argued, it could be seen that it was in the interest of each species to adjust its reproductive rate to match the food supply available to it, and to balance the forces of predation upon it. In this way, Forbes imagined, there would develop a common interest between a species and its enemies. A predator must not diminish its prey below a certain level, otherwise it would threaten its own survival. Similarly, a prey species must maintain a reproductive rate to compensate for the destruction of individuals through predation. Natural selection would promote the survival of species that had adjusted their reproductive rates to serve these common interests, eliminating those that were maladjusted. The key to survival was shrewd economizing, just as it was in the human community. As Forbes wrote in his 1887 essay "The Lake as a Microcosm": "Just as certainly as the thrifty business man who lives within his income will finally dispossess his shiftless competitor who can never pay his debts, the well-adjusted aquatic animal will in time crowd out its poorly-adjusted competitors for food and for the various goods of life. Consequently we may believe that in the long run and as a general rule those species which have survived, are those which have reached a fairly close adjustment in this particular."[25]

Evidence of perfect adjustment could be seen by perfect stability in numbers from one year to the next. But perfect adjustment was seldom achieved: more often, species would oscillate in numbers as they acted and reacted upon each other. Any major disruption of the balance would create severe oscillations, indicating maladjustments of the population. It was these oscillations that the scientist had to learn to control, for such drastic changes would nearly always be injurious in nature. Despite nature's fierceness, despite the intensity of the struggle for existence, natural selection was basically a beneficent force, because it tended always to restore a healthy equilibrium, conducive to the maximum common good. And, Forbes could not help but note, if hostile and indifferent nature should have produced an order that was beneficent and harmonious, how much more might be expected in human affairs, where "the spontaneous adjustments of nature are aided by intelligent effort, by sympathy, and by self-sacrifice?"[26]

Filtered through Forbes's liberal Protestantism, Darwin's struggle for existence and Spencer's balance of forces combined to produce a benign and thrifty image of nature. Harmony was maintained by the checks and balances which were expressed through the complex predatory and competitive relations of the community. It was the task of the scientific natural historian to understand how those relations worked, and how the community was structured and maintained, so as to know how to manage nature with prudence and foresight in man's best interests. Such was the ecological perspective that Forbes elaborated, both in its practical and philosophical aspects, from the 1880s onward.

At the time that Forbes was developing his point of view, the word "ecology" was still not in general use, but this deficiency was soon to be remedied. By the 1890s the distinction between the two senses of physiology, one narrow and the other broad, had become sharp enough for botanists to feel the need for a new term to describe their interests in the broader style. In 1893 a small group of American botanists, meeting in Madison, Wisconsin, decided to appropriate Haeckel's word for this purpose. Eugene Cittadino has noted that an ecologist was understood to be basically an outdoor physiologist, a person who "made measurements, recorded the responses of plants, and tried to show functional relationships between the structure and responses of individual plants, or groups of plants, and various environmental factors."[27] Zoologists were not long in following suit, adopting the word "ecology" for their science as well; one of the first to do so being Stephen Forbes, who referred to himself as an "oecologist" as early as 1894.[28]

Although ecology was first distinguished as a branch of physiology, there were some differences of opinion over how nature might best be subdivided for ecological study. For some, ecology remained focused on the individual organism, but for others it was apparent that ecology must include the study of larger, integrated groups of species. Plant ecologists took the plant community to be their natural unit of study, while the animal ecologists looked at the "association," "biocoenosis," or "living community."[29] In general, ecologists tried to go beyond the simple description of a community in order to understand the processes giving that community its coherent structure and guiding its change over time.

In analyzing the processes at work in these larger natural units, ecologists were easily drawn to organismic analogies. Given their physiological interests, this frame of reference was a natural one to adopt. Forbes, for instance, spoke of the "sensibility" of the organic complex, by which he meant that whatever affected any species belonging to the community would ultimately have an influence on the whole assemblage.[30] He did not press the organismic analogy, but used it to draw attention to the functional relations between the inhabitants in well-defined communities, such as lakes.

A more literal use of the organismic viewpoint was made by Frederic E. Clements, the leading plant ecologist in America in the early twentieth century. Clements had also been influenced by the ideas of Herbert Spencer in his youth, as well as by the German plant geographers. He perceived the community to be a complex organism which grew, matured, and died like an individual organism.[31] This frame of reference led him to describe plant succession as a process of continual interaction between the habitat and the life-forms of the community. The habitat and the populations acted and reacted upon one another in an alternating cause and effect sequence, until finally an equilibrium state, the climax community, was reached. Succession was directed by physiological processes of competition occurring between similar plants with the same physiological needs. Species of trees were said to compete sharply when together, because they were similar in form, while the relation of shrubs to trees was thought to be one of subordination and dominance rather than competition.[32] Successional development was accomplished through the gradual reduction of competition; that is, the regular outcome of competition was to set up more or less stable dominance hierarchies among the plants of the community, as it moved toward the final climax stage. Therefore, an increase in the diversity of the community was thought to lower the total amount of competition and to produce a stable state. In keeping with the idea of the complex organism, Clements considered succession to be always progressive, leading to one climax and never going backward.

Clement's organismic language drew attention to the continual action and reaction of organism and habitat, or of complex organism and environment. Even among those not attracted by the organismic metaphor, the language of action and reaction pervaded biology at the turn of the century. As Garland E. Allen has observed, biology at the time was moving away from the static morphological view of nature to the dynamic experimental view.[33] Along with that shift came a better appreciation of the complex relations prevailing between organisms and environment. The completely deterministic, machinelike image of organisms put forth by turn-of-the century mechanists, such as Jacques Loeb, was starting to give way to a less rigid interpretation suggested by such biologists as Herbert Spencer Jennings. Jennings called attention to the organism as a "complex of many processes, of chemical change, or growth, and of movement," proceeding with a certain energy that could flow and overflow in varied directions.[34] The study of life was the study of regulation and of how the animal kept itself regulated both physiologically and behaviorally. This holistic materialism, as Allen has called it, continued to gain adherents in the first two decades of the century, finding philosophical expression in Alfred North Whitehead's *Science and the Modern World*, his 1925 Lowell Lectures at Harvard.[35]

Ecologists were also moved by these currents. Ecology being a new science, it was important to set it on the right track at the start, to provide some principles which would integrate and organize the huge mass of ecological facts and bring ecology into line with the progressive ideas of the experimental sciences. Such was the feeling of Charles Christopher Adams, who from 1896 to 1898 was working as an entomological assistant at the Illinois State Laboratory of Natural History, where Forbes was still director. Adams appreciated the integrated approach of Forbes, but a more important influence was Henry Chandler Cowles, whom Adams met when he went to the University of Chicago in 1900. Cowles, who imprinted his point of view on many young ecologists of this generation, was starting as an instructor in botany when Adams arrived, and he had just finished a study of the physiographic ecology of Chicago and its vicinity, published in 1901.[36] Trained in geology, Cowles imparted to Adams a particularly geological perspective, a concern with the orderly succession of communities by "imperceptible gradation," just as there was an "order of succession of topographic forms in the changing landscape."[37]

From 1903 to 1907 Adams was involved in museum work at the universities of Michigan and Cincinnati. Receiving his Ph.D. from Chicago in 1908, he began work on a cooperative project in ecology between the Illinois State Laboratory of Natural History and E. N. Transeau and T. L. Hankinson of the Ecological Survey Committee of the Illinois Academy of Science. He had long been concerned about the abundance of facts and the poverty of theory in ecology, and as he worked up his section of the report for this project, he developed his ideas for an approach to ecology that would help to guide its orderly development. His ideas culminated in his *Guide to the Study of Animal Ecology* (1913), an annotated bibliography with introductory essays outlining his point of view, which he called the "dynamic" or "process" method of ecology. Over the next few years, he elaborated the method, drawing from appropriate ideas in the allied sciences as they came to his attention. At the same time he worked to establish ecology as a discipline, being one of the founding members of the Ecological Society of America in 1915, and its president in 1923.[38]

The dynamic-process method incorporated many of the current methods of modern experimental science. The key ideas of this new approach were energy and system, a system being characterized by the ability to respond to disturbance and to maintain a state of equilibrium. The entire world was a hierarchy of systems: "From electrons, atoms, molecules, chemical compounds, colloids, cells, tissues, organs, individuals, and culminating in the community and association, is seen in each a dynamic center or microcosm, about which revolves other systems, in turn revolving as a part of a larger system in ever widening expansion, each in turn subordinated to a higher order of dominance, the culmination of interacting systems."[39] As with Forbes and Clements, the shadow of Herbert Spencer loomed not far in the distance, for Adams felt that no one had put forth a better discussion of the dynamic concepts of systems than Spencer. But Adams was reacting more immediately to other ideas closer at hand, principally ideas derived from physical chemistry.

Physical chemistry, a relatively new branch of science that emerged in the late nineteenth century, was concerned with the description of chemical systems using the methods of physics, especially thermodynamics. The systems approach was useful because it looked for general laws which could be applied to systems that appeared to be quite different in composition: thus it provided a basis for comparative analysis of systems. One general law derived from thermodynamics was a version of Newton's third law of action and reaction, formulated by Le Châtelier in the 1880s to describe the conditions for equilibrium in chemical systems. Le Châtelier's principle was afterward reformulated in a variety of ways, many of them more comprehensive than his original version. The concept of equilibrium, often expressed in various formulas that resembled the formulas of the physical chemists, enjoyed considerable vogue in the early twentieth century, pervading both biological and social thought.[40] In Adams's case, he chose a very broad statement of the principle, given by W. D. Bancroft in 1911: "A system tends to change so as to minimize an external disturbance."[41]

With the help of universal laws of this kind, Adams felt, ecology could be organized around problems concerning the maintenance of equilibrium in more complicated organic systems. What he did was to combine the concept of the complex, interacting organism, as put forth by H. S. Jennings and others, with the central ideas of physical chemistry: system, energy flow, action and reaction. From this he hoped to build a set of ecological principles that would put some order into this young and confused field. He was encouraged in his attempts to join biology with the physical and chemical sciences by Lawrence J. Henderson's recent book, *The Fitness of the Environment* (1913), which neatly set out the importance of physicochemical processes for the understanding of biological ones. Henderson was also a keen advocate of equilibrium analysis and its possible applications to the social sciences.[42]

The dynamic-process viewpoint began by seeing the organism as an energy transformer, and from there investigated the processes of energy transformation that accompanied the changes in and out of equilibrium states. The emphasis on "processes" caused Adams to perceive evolution itself as a process that included all the changes taking place in the organism and the environmental implications of those changes. Evolution was taken to be a general concept, along the lines of the concept of "metabolism." This broadening of the meaning of evolution led to a de-emphasis of the end products of evolution, the species, in favor of a search for the processes themselves: the process of living, the process of evolution, was thought to be of more importance than the products of evolution, such as species and varieties. The emphasis changed from a study of forms to a study of functions: "The products must be subordinated to the agencies and processes, because the laws of change are in reality the object sought."[43]

Although we can see the value of making the point that dynamic relations are part of evolutionary changes, the shift in attention from product to process looks exaggerated to our eyes. It should be kept in mind, however, that far from being an isolated notion invented by Adams, this was a pervasive scientific point of view. Thorstein Veblen drew attention to it in the early 1900s by noting that modern science was becoming "substantially a theory of the process of consecutive change, realized to be self-continuing or self-propagating and to have no final term."[44] In his idiosyncratic style, Veblen attributed this evolution of science to the domination of the machine process in the modern age, where industry and industrial processes had come to shape human thought, such that people had started to think "in the terms in which the technological processes act."[45] Veblen noted the signs of this habit of thought in the physical and chemical sciences, but it was already filtering into the biological sciences as well.

Physical chemistry provided only part of the basis for dynamic-process ecology, however. The other part came from the geological sciences, with which ecology was also allied. The emphasis on processes rather than products reflected a geological perspective, one which Adams traced correctly back to Charles Lyell, but which had more recently been enunciated in the geological sciences, and from there through Henry C. Cowles had been applied to plant ecology. Adams drew explicitly upon this geological frame of reference, comparing the processes an animal undergoes to the geological processes of weathering. Just as running water is an agent of the process of erosion, so an organism is "an agent which expends physical and chemical energy, producing stress and exerting pressure and expending energy on other substances." Adams continued, "An animal, by the process of predation runs down another animal and devours it, by its *process* of digestion dissolves it, and by the process of assimilation makes muscle, bone, feathers or fur out of it, and these are all products of its activity. The process of response is here strictly comparable to the process of erosion of running water, and their products are similarly comparable."[46]

The dynamic-process method provided not only a way of understanding individual organisms as agents of change in a constantly changing world, but also a way of understanding the larger ecological units. Without using the organismic metaphor of Clements, or the "sensibility" of Forbes, Adams expressed the same idea of the organized community responding as a whole to stimuli: "The association, as a whole, is thus in continuous process of bombardment and response from every possible angle, and just as the individual animal is stimulated and responds, so all the members of any association are stimulated and respond in a similar manner. It is by this form of activity that animals not only maintain themselves but exert a radiating influence."[47]

Adams's intention in setting out this method was not to present a fully formed body of ecological theory. Indeed, his ideas were mere sketches, outlines of possible approaches, and questions which might be usefully asked. His more important goal was to suggest that ecologists can and should make use of hypotheses to guide their research; that a wholly inductive method would not be fruitful in the long run. He realized that his systems approach contained many assumptions which had not been tested, and could on that basis be faulted. But if it directed attention to important experiments, its function would have been served. Adopting the point of view of the early nineteenth-century Scottish geologists, Adams turned their philosophical insights to the allied science of ecology. It was thoroughly appropriate that he should choose to quote from John Playfair's *Illustrations of the Huttonian Theory of the Earth* (1802) to point out the importance of a good balance between hypothesis and observation:

> Though a man may begin to observe without any hypothesis, he cannot continue long without seeing some general conclusion arise; and to this nascent theory it is his business to attend, because, by seeking either to verify or to disprove it, he is led to new experiments, or new observations. . . . Thus theory and observation mutually assist one another; and the spirit of system, against which there are so many and such just complaints, appears, nevertheless, as the animating principle of inductive investigation. The business of

sound philosophy is not to extinguish this spirit, but to restrain and direct its efforts.[48]

Ecology had hardly emerged when already ecologists were concerned with the problem of how best to impose a unified theoretical structure on the facts of nature. This was a difficult task in a discipline that embraced such a wide range of activities. In its most general sense, ecology was the study of the day-to-day processes that make up the struggle for existence. That meant first of all the study of how the individual was suited to the conditions under which it had evolved and now lived. This was a morphological problem; the study of adaptation, of how structures could be better understood by looking to the wider ecological context. It was also a physiological problem; a study of how organisms responded to the environment, how they changed the environment, and how they responded to the changes caused by other organisms. Here the ecologist was an outdoor physiologist.

Ecology was also the study of how groups of organisms lived together; how the combined effects of many species living and interacting together produced what appeared to be stable structures: communities, societies, associations, biocoenoses. Here the ecologist was also a natural historian, part taxonomist, part census taker, laboriously collecting, counting, measuring, trying to determine what species made up the community; what the individual life histories were of its inhabitants; how the food web was constructed; how organisms competed, or how they cooperated.

Finally, having identified these larger units, the communities, and having seen that they preserved a relatively stable structure, the ecologist wanted to know how the communities as a whole changed over time. What were the processes of succession, the gradual, almost imperceptible changes that transformed the surface of nature? Here the ecologist was geographer and geologist; mapping and classifying communities, setting down the laws governing the orderly sequence of nature.

Knowing nature's orderly sequence meant being able to manage nature wisely. Ecology justified its grounds for existence by insisting on its importance as an applied science: many of the central ideas, concepts, and methods of ecology arose from attempts to solve practical problems in agriculture, forestry, and fish culture. The benefits of sound management could easily be converted to dollars and cents. Forbes estimated in 1880 that if the result of his investigations showed how to increase the efficiency of birds as insect predators by as much as 1 percent, the savings to the agriculture of Illinois would amount to $66,000 per year, "equivalent to the addition of over one and one-half million dollars to the permanent value of our property."[49] And, as Forbes shrewdly pointed out, it was the promise of solid practical benefits which mostly justified the laborious field work required in ecology. Adams took the justification for an applied science one step further, arguing by the 1930s for an ecological approach to human problems, one that would draw upon the records of history and archeology, as well as the methods of sociology, economics, and urban geography.[50] Ecology would be the applied side of biology, borrowing from the methods of all the allied natural and social sciences as needed.

While it certainly helped the growth of ecology to be so clearly identified as an applied science, it also meant that less practical projects received less attention. Already in 1889 Forbes was noticing that the work in "general zoology" at his laboratory had fallen off, as a result of changes of assistants, and that this was indirectly the result of the organization of the agricultural experiment stations, formed after the passage of the Hatch Act in 1887.[51] Greater attention was to be devoted to economic entomology, less to more general problems in zoology. This balance between general and applied research carried implications for the way ecologists would receive new theoretical ideas in their discipline. Practical biologists are closely tied to a given biological reality: they want to predict and to control a specific segment of nature. Their concern with practical problems makes them correspondingly skeptical of theoretical approaches which diverge too far from the reality they know.

But even a small portion of reality was exceedingly complex. If ecology was to lay bare the orderly sequence of nature, it had to begin by simplifying nature. Paradoxically, the way to understand complexity was by being unabashedly simplistic at the start; hence the appeal of Herbert Spencer's ideas. Ecology needed some organizing theory, but it was less clear how best to impose some structure on nature in a way that would prove fruitful. Adams felt the dynamic-process viewpoint was valuable because it emphasized causal relations. Others would later offer their own strategies, and ecology would find itself becoming increasingly mathematical in the following decades.

Already, however, there were intimations in the words of Adams that no theory would ever reach the full truth. Approximations could be made, but there would always be deficiencies in the accuracy of their descriptions. Ecology would involve a continuous process of hypothesizing, experimenting, observing. There would be no final term, no complete answer, only the gradual unfolding of an immensely richer and more complex world.

2

The World Engine

In 1925, Charles Adams's eye was diverted by a new book called *Elements of Physical Biology*.[1] The book was a study of the dynamic processes of nature, touching on the very issues that Adams had been pressing upon the ecological community for the past two decades. Its author was not an ecologist, nor even a biologist, but a physical chemist by training, named Alfred James Lotka. For over two decades, Lotka had been working quietly away and had finally gathered his thoughts into this unusual book, whose title scarcely revealed its grand scope. The organic world and its inorganic parts were regarded as a single system, a term implying not only that each component part was linked to every other part in some way, but that is was impossible to understand the working of any part without an understanding of the whole.

The analysis of the entire world system required new principles which Lotka borrowed freely from physics: physical biology was the application of physical principles to biological systems. He distinguished his approach from the companion field of biophysics, which involved a similar application of physics to biology, but where the system under study was the individual organism and the problems were of a morphological and physiological nature. Lotka's system was the whole world: the problems therefore involved aggregates of organisms, studied in the same way that a chemist might study aggregates of molecules. Lotka's inspiration came from physical chemistry, and the language he used, with references to the "kinetics of evolution," "statics" and "dynamics" of living systems, as well as the title itself, reflected this influence.

Adams had been struck by the ecological tone of much of Lotka's discussion. In the course of outlining his method, Lotka had delved into such topics as food webs, the water cycle, and the carbon dioxide, nitrogen, and phosphorus cycles, all presented with a good selection of supporting quantitative data. He had summarized much of the literature from aquatic biology and had suggested that the term "general demology" be used for the quantitative study of organisms living in mutual dependence. The book as a whole emphasized the importance of studying the relations between organisms from the energetic point of view. Moreover, where

Adams's discussions had been sketchy and imprecise, Lotka had tried to express his ideas as exactly as possible with mathematics.

Adams also observed that, although this book had much to offer ecologists, its author did not seem to be in touch with the ecological community. He wrote to Lotka, suggesting that he consider a review in *Ecology*.[2] The review was never written, but Lotka did take Adams's advice and joined the Ecological Society of America in 1925. Thus began a connection between ecology and a reclusive mathematician who had never intended to write an ecological work, but had tried instead to start a whole new discipline in science—physical biology—meant to be analogous to physical chemistry. Lotka failed to found his new discipline, but by the time his book was reprinted in 1956 as *Elements of Mathematical Biology*, it was recognized as an ecological classic, though he never identified with the community that had responded earliest to his ideas. His story is of interest not only for the novelty of his ideas and methods, but because he was an example of a rare creature in modern science; a scientific amateur, unconnected to an academic institution, struggling to get his ideas known by making connections wherever he could.

Lotka's principal reputation in science is based on the work he did after finishing his book, when he gave up his amateur status to become a professional demographer. The recognition afforded him as a demographer contrasts with the curiously marginal position given his book. The *Elements* itself is usually cited in ecological literature for two reasons. The first is its clear exposition of the systems approach which later had some influence on ecosystem ecology as developed by Eugene P. Odum and Howard T. Odum in the 1950s.[3] However, two other ecologists, Arthur G. Tansley and Raymond L. Lindeman, used the concept of the ecosystem before the Odums without in any way being influenced by Lotka.[4] His treatment of the world as a system also coincided with the biogeochemical approach to ecology being developed by Vladimir I. Vernadsky in Russia.[5] Lotka immediately recognized the connection between his work and Vernadsky's, although the latter's research had not been published in time to include more than passing mention of it in the *Elements*. The second reason why ecologists cite his book is the section on predator-prey oscillations, which occupies but a few pages. This mathematical treatment of population interactions became widely known mostly through the work of Vito Volterra, a mathematical physicist of renown, who happened to come up with the same result just as Lotka's book was published.

Apart from the ecological import of Lotka's ideas, his book also anticipated Ludwig von Bertalanffy's development of general system theory in the 1950s.[6] Bertalanffy rather ungenerously downplayed the similarities between his method and Lotka's, but Lotka had clearly set down the basic procedure of systems analysis first: he just did not apply it as extensively as

Bertalanffy did. In economics and the managerial sciences, appreciative references to Lotka may be found in the works of Paul A. Samuelson, Henry Schultz, and Herbert A. Simon.[7] Simon summed up Lotka's impact by calling him an imaginative forerunner who "creates plans of exploration that he can only partly execute," but who guides his successors by "posing for them the crucial questions they must answer, and disclosing more or less clearly the directions in which the answers lie."[8] In ecology, his impact was of a similar kind: his ideas were intriguing in their suggestions of how mathematics might be used in ecological contexts, but most were not well enough worked out, and too biologically naive, to have served as a direct stimulus for action.

Alfred James Lotka, 1880–1949
Photograph courtesy of Metropolitan Insurance Companies

For all Lotka's achievements as a forerunner, his book and his program are oddly backward looking. The book evinces a distinctly nineteenth-century sensibility in its grand, synthetic approach to science. It is in many respects an elaborate extension of the synthetic philosophy of Herbert Spencer, welded to a modern, mathematical frame, and updated to include the concerns with science and technology that were preoccupying the twentieth-century mind. Larded as it is with contemporary philosophical quotations, it provides an overview of the ideas current in the first two decades of the century. It also provides a fascinating glimpse of the amateur's imagination, unfettered by the disciplinary constraints of professional science. Lotka drew freely on all the sciences: physics, chemistry, biology, economics, and the social sciences. Its broad scope puts his book into a class by itself and accounts for its failure to find the audience that Lotka sought. There is no denying that the book was a failure; but it was an interesting failure, one which points up the necessity of establishing an institutional basis if one's purpose is to advance a new scientific discipline. Before looking at the areas of population ecology where Lotka did have an impact, therefore, it is worth spending time on his program as a whole to understand how he failed, and why.

The Mole and the Eagle

Lotka was born of American parents in Lemberg, Austria (now Lvov, the Ukraine), in 1880, and was educated in Germany and France. He took a bachelor degree in physics and chemistry in 1901 at the University of Birmingham, England, where he came under the spell of John Henry Poynting, himself a student of James Clerk Maxwell. He was inspired by Poynting's teaching and later dedicated *Elements of Physical Biology* to the memory of his old teacher.

From Birmingham he traveled immediately to Leipzig to pursue his studies. Leipzig at that time was a center of the new field of physical chemistry, a union of physics and chemistry emphasizing the use of thermodynamic principles in chemical systems. The founding father of the energetic approach to chemistry, Friedrich Wilhelm Ostwald, was then the director of the new Physical-Chemistry Institute which had opened in Leipzig in 1897.[9] During the year that Lotka was there (1901–1902) Ostwald delivered a series of lectures, later published under the title *Vorlesungen über Naturphilosophie*,[10] in which he propounded the idea that energy was the central organizing concept of the physical and biological sciences. It was one of these lectures that started Lotka on the train of thought that culminated in the *Elements*.

After his year in Leipzig, Lotka went to the United States, settling in New York where he worked as an assistant chemist for General Chemical Company. The next twenty years were mostly taken up by the burden of earning a living. He left General Chemical in 1908 to become a graduate student and assistant in physics at Cornell University, where he picked up a master's degree in 1909. The year at Cornell was followed by a short stint as an examiner in a patent office, then a job as assistant physicist in the U. S. Bureau of Standards from 1909 to 1911, and finally a period spent as an editor at *Scientific American Supplement* from 1911 to 1914. During the war years Lotka again worked for General Chemical Company, where he was involved in research on the fixation of atmospheric nitrogen.[11]

Despite the fact that Lotka was never fully satisfied with these jobs, he did not pursue an academic career. Instead he followed his intellectual interests in his spare time, publishing a series of articles and notes, beginning in 1907, in a variety of academic journals. Twelve of these publications formed the basis of his submission for a D. Sc. from the University of Birmingham, which he received in 1912.[12] Most of the articles dealt with the application of physical principles to biology, but the range of topics was wide, stretching from demographic studies of age distribution and the relation between birthrates and deathrates, to quantitative studies in epidemiology, to a study of evolution from the standpoint of physics. A few were only short notes on miscellaneous subjects, such as his paper on the "Construction of Conic Sections by Paper Folding,"[13] but the majority pertained in some way to the subject of physical biology.

This early work received scant notice. His occasional hints that the themes explored in each article were part of a broader program of investigation did nothing to illuminate its purpose. Lotka labored on, building his program piece by piece; but he was like a mole, anxious to soar yet equipped only to tunnel deeper underground. His schemes and scratches had brought him no closer to recognition by 1920, when his small movements caught the eye of an eagle, which swooped down to raise him up in its large but friendly talons. The eagle was Raymond Pearl of the Johns Hopkins University.

Pearl was already a man of considerable influence in science. He had served during the war as chief statistician for Herbert Hoover's Food Administration Program, and he had been elected to the National Academy of Sciences at the relatively young age of thirty-seven. He had recently moved to Johns Hopkins as the first Professor of Biometry and Vital Statistics in the new School of Hygiene and Public Health. He was a keen advocate of statistics in the biological and medical sciences, and was trying to introduce the techniques of demography into the study of animal populations. Lotka probably did not know of Pearl's population interests at the time, but Pearl was sharp enough to notice that there were certain points of contact between Lotka's mathematical and demographic approach to biological systems and his own newly acquired interests in population biology. In 1920 Pearl sent a short article by Lotka, entitled "Analytical Note on Certain Rhythmic Relations in Organic Systems,"[14] to the National Academy of Sciences. In the spring of 1921 he invited Lotka to Baltimore for a few days to give some lectures on his research and to explore the possibilities of establishing a permanent connection with Pearl's laboratory at Johns Hopkins. "I think I know when a man has genius," he soon afterward wrote to Lotka, "and I also, I believe, fully realize how any thing that a university can do to aid the work of such a man is but meagre compensation for what he does for mankind."[15] Lotka was heartened by Pearl's continued support, especially as the small audiences at his Baltimore talks had left him with a sense of failure:

> The kind of things which you say regarding my work stir in me emotions which I shall not attempt to describe. You may form some conception of them if you reflect that it is now nearly twenty years since I have started my investigations; and during all that time there has been no dearth of discouragements in varied assortment, of recognition hardly a trace. It is not that a man works for the sake of recognition, but, after all, one's courage has its limits.[16]

Pearl's offer, however, was only to open the facilities of his laboratory for Lotka's use. No salary was involved. Lotka hesitated to commit himself when he was unsure of how much time he would be able to spend in Baltimore. A plan was worked out whereby Lotka would live in New York and freely use the Baltimore facilities whenever convenient. The only condition was that all of Lotka's articles would thereafter appear in the numbered series of contributions from the Department of Biometry and Vital Statistics at Johns Hopkins. After some discussion about what designation would be appropriate for such a connection, the faculty finally settled on Fellow by Courtesy, an honorary title which carried no formal responsibilities. The fellowship was duly passed by the board of trustees in June 1921, and that summer Lotka began to address his papers from Baltimore.[17]

Around the same time he also began to think seriously about gathering his ideas into book form. Pearl had suggested that a volume connecting all of Lotka's work together would be desirable "so that it will make the impression that it ought to on the scientific public, which it cannot do in the series of scattered papers."[18] Lotka entirely concurred. "Besides," he wrote to Pearl, "I feel that one of the principal uses of the work lies in the comprehensive outlook gained and the systematic 'orderly array' into which it throws a large number of facts thus found to stand in relation to each other."[19] The problem was to find financial support for the period needed to write the book, which Lotka estimated would occupy at least two academic sessions.

One possibility was the Rockefeller Institute for Medical Research, whose operations had both impressed and amused Lotka on a visit to Alexis Carrel's tissue-culture laboratory in the summer of 1921:

> I saw a young lady making chicken hash in the most heartless but thoroughly scientific manner. I saw a little piece of chicken heart foolishly throbbing after all occasion for throbs had long ceased. I saw a thirteen year old dog rejuvenated by the infusion of younger blood. It was hardly converted into a giddy young thing, but they told me there was still marked improvement even after twelve months.[20]

Lotka was unable to speak to Carrel, the head of the laboratory, but he broached the subject of a possible connection to the institute with a member of the scientific staff, A. E. Ebeling, who seemed to respond favorably. Lotka's work was not of a medical nature, but he was working at the time on a lengthy mathematical study of malaria, which he hoped would demonstrate the relevance of his research to medical problems. He planned to apply to the Rockefeller Institute as soon as this study was completed.[21]

Pearl was rather astonished to hear of these new plans, but he assured Lotka that he could rely on full backing from the Johns Hopkins. However, he expressed doubts that Lotka's scheme would meet with much success:

> Dr. Abeling's [*sic*] position is a very subordinate one and what he says or recommends is not likely to have very great weight with the governing powers. I do not say this to discourage you, but simply because I have taken part in a number of attempts to get other very able men attached to the Rockefeller Institute, and have found the path beset with difficulties.[22]

Pearl suggested that Lotka contact Alexis Carrel personally, but added that he would much prefer to have him at his own laboratory. He promised to speak to William Henry Welch about the possibility of making Lotka an attractive offer. After weighing Pearl's cautious remarks, Lotka decided to let the Rockefeller plan rest for a while, and he devoted his attention to the malaria paper.[23]

Pearl had still not made a specific offer when, in early 1922, Lotka noticed in the *Johns Hopkins Circular* that a fellowship of $1,000 was being offered by Pearl's department.[24] He debated whether or not to apply, for it meant giving up a salary of $3,000 and perhaps losing the opportunity to bargain for a higher stipend elsewhere. Pearl, who had earlier urged Lotka to maintain his connection to the Johns Hopkins, now raised the objection that Lotka's position as Fellow by Courtesy might make it psychologically awkward to award him a less distinguished fellowship: "A fellowship of Courtesy is regarded here as a very high honor and belonging properly to men of a distinctly higher rank of attainment than those to whom we give our regular fellowships which carry a stipend."[25] After consultation with other members of the board of trustees, however, Pearl was able unofficially to assure Lotka that the fellowship would indeed be granted and encouraged him to apply.[26]

By this time, Lotka had become increasingly anxious to write his book because he feared that someone would anticipate his ideas. The perception of nature as one gigantic whole, though treated by Lotka in a unique way, was a common point of view in several fields in the 1920s. This holistic or organismic perspective paralleled Lotka's own in its suggestion that not only did the existence of the whole depend on the orderly cooperation and interdependence of its parts, but that the whole also exercised a measure of determinative control over its parts.[27] Holistic ideas which had their origins in physiology were being extended to psychological studies dealing with sense perception, and the nature of consciousness and learning.[28] These were all questions that Lotka had touched on in his writings. Finally, the holistic worldview aimed at a scientific expression of a complete picture of the world, thereby putting an end to what was seen to be the inappropriate separation of science from aesthetic and ethical experiences. Lotka hoped to do the same.

Another development which he felt as a threat to his priority was the growing body of literature in biophysics. D'Arcy Wentworth Thompson's two-volume study of the application of mathematics to problems of growth and morphology, *On Growth and Form*, had appeared in 1917.[29] More recently, Alexander Forbes of the Harvard Medical School had published an appeal in 1920 for greater coordination between the work of physicists and biologists in the study of physiological problems.[30] A new book on biophysics by David Burns,[31] though entirely physiological in its emphasis and therefore quite different from Lotka's approach, was viewed with alarm:

> The time is ripe. I see many signs of the fact. Have you seen a recently published book by Burns and Paton, An Introduction to Biophysics? It also dug a spur into my side—not that there is anything to get excited over, the book would hardly be described as either inspired or inspiring. But it shows the undercurrent, which one of these days must break through to the surface.[32]

These currents carried him to a decision. In 1922, at the age of forty-one, he accepted the fellowship at Johns Hopkins and moved to Baltimore, where he spent the next two years writing *Elements of Physical Biology*.

The Program of Physical Biology

The *Elements of Physical Biology*, conceived in 1902 but completed more than twenty years later, expressed the synthetic spirit of the nineteenth-century thought that had prompted it. Its message was the intricate

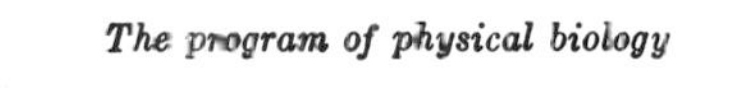

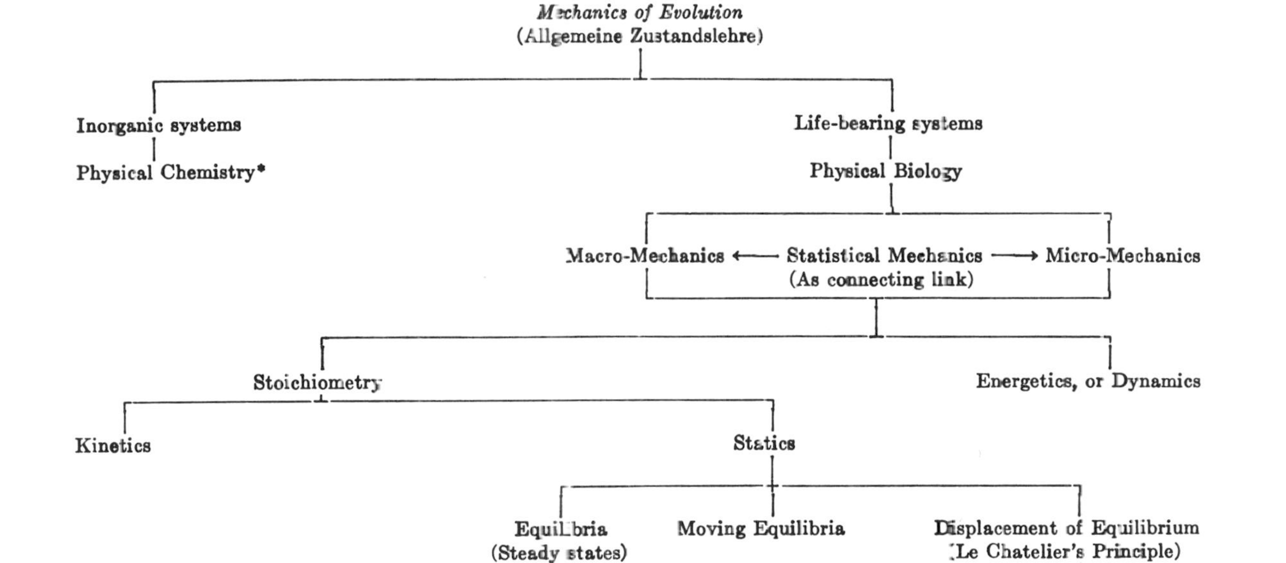

* This term may here be taken to include the treatment of physical *Change of State* (evaporation, fusion, etc.).

FIGURE 2.1. Lotka's diagram showing the different parts of his program in physical biology. (From A. J. Lotka, *Elements of Mathematical Biology*, New York: Dover Press, 1956, p. 53.)

connection between all things, which meant not only the interrelation of the animals, plants, and inorganic matter of the world, but of the branches of science which dealt with those relationships at all levels (Figure 2.1). The basic problem providing the framework for the program of physical biology was evolution: not, however, the evolution of individual species, but the evolution of the whole system. If life were a drama, for Lotka there could be no distinction between stage, scenery, and actors; all were partners in the same "intimate comedy." "And if we would catch the spirit of the piece," he wrote, "our attention must not all be absorbed in the characters alone, but must be extended also to the scene, of which they are born, on which they play their part, and with which, in a little while, they merge again."[33]

The contemplation of this "world evolution" entailed the analysis of the complex relationships that bound the world's inhabitants together. The field was thus opened to a wide range of problems: population growth, predator-prey interactions, nutrient cycling, animal behavior, to name but a few examples, all found their places in the program. For this reason the book had a strongly ecological flavor, although Lotka himself did not perceive at first the connection between the growing science of ecology and his own book. It is a measure of his estrangement from the academic community that, as late as 1927, he could still express disappointment that biologists, rather than physicists, had been the ones to respond most to his book.[34]

The viewpoint that Lotka presented in the *Elements* was based on the proposition that a biological system could be analyzed in the same way that a physical chemist would analyze a chemical system. All of the processes occurring in these systems could be reduced to two fundamental kinds of changes: those involving exchanges of matter between the components of the system, and those involving exchanges of energy. These two types of transfers were really aspects of the same thing: transfers of matter involved transfers of energy. The distinction between the two, he thought, was merely a practical one: sometimes it was convenient to view an exchange as a transfer of matter, other times it might be easier to examine it in terms of energy. In the chemical system, the components were the molecules, and the exchanges of matter and energy were accomplished through the various processes of chemical reaction. In the biological system, the components were the organisms and the inorganic raw materials, and the exchanges took place through the complicated web of food relationships, growth, and reproduction. The analysis of these relationships in a mathematical, quantitative manner constituted the program of physical biology.

It is difficult to determine just when it occurred to Lotka that his program might be formulated in the context of a new discipline called physical biology. In the chapter in which he outlined his program, he began with a quotation from a journalistic piece written in 1915 by W. Porstmann for the German journal *Prometheus*, a weekly devoted to various topics in trade, industry, and science.[35] Porstmann had argued for the need of a new science of physical zoology and botany, somewhat along the lines of physical chemistry, but otherwise his discussion had little relevance to Lotka's approach to the subject. It is possible that this article gave Lotka the disciplinary framework in which to fit the ideas he had been thinking about for several years.

In any case, the similarity between physical-chemical and biological systems had first struck Lotka much earlier, while he was attending Friedrich Wilhelm Ostwald's lectures in Leipzig in 1901. In one of these lectures, Ostwald compared the growth of a bacterial colony to the formation of crystals in a supersaturated liquid.[36] Such a liquid existed in what he called a "meta-stable" state, that is, it was in equilibrium until disturbed by the addition of a crystal which would act as a "seed" for the formation of more crystals, until a second equilibrium of concentration was attained. The whole process was accompanied by energy changes within the system. In a similar manner, the bacterium "seed" in a nutrient broth grew by extracting solid matter from the surrounding liquid, a process accompanied by energy changes in the living colony. Ostwald's analogy, introduced in passing at the end of a lecture, was only a heuristic device used to illustrate how biological processes might be understood by reference to inorganic processes. As Lotka later recalled, it was this comparison which had acted as the "trigger" for the train of thought that led eventually to the *Elements*.[37] The important difference was that Lotka's argument rested not on superficial resemblance, but on the demonstration of a true identity between physical-chemical and biological systems. This demonstration required an elaboration of Ostwald's sketch.

The most obvious difference between chemical and biological systems was the degree of complexity of the systems themselves. Biological systems were heterogeneous: local environmental conditions varied both in space and in time, and the components themselves, the organisms, could also vary in structure over time. In the biological system, therefore, one had to consider the *structure* of the system itself, its geometrical configuration, in order to understand the changes going on within it. The geography of the environment, for instance, or the distribution of organisms within it, might well determine how and when they could interact.

Chemical systems by comparison were fairly homogeneous with respect to environmental conditions and consisted of a few components that combined in predictable ways. In these systems, geometrical considerations could be ignored by limiting oneself to an appropriately simple, homogeneous system. For Lotka, however, this difference was only a practical one adopted by a new science, physical chemistry, which necessarily began with the simplest cases. In theory it might be possible to take complicated structural features into account when analyzing chemical systems, in which case the similarity between chemical and biological systems would be more apparent. It was in this type of theoretical, structured chemical system that Lotka perceived the similarity to biological systems.[38]

Although it might seem puzzling that Lotka would go to such lengths to posit a hypothetical chemical system as the basis for such a comparison, from his point of view it was necessary to do so in order to forestall the accusation that his subsequent analysis rested on superficial analogy alone. The success of Lotka's program depended on the reader's willingness to perceive the similarity between these systems as one based on actual identity in type.[39] An incidental similarity based on loose analogy was not sufficient to justify the application of physical laws and techniques to the analysis of biological systems on the scale that Lotka envisaged them. Though he may be said to have stretched the imagination at this early stage, his characteristic concern about the use of analogy as a basis for the interpretation of biological events was valid, and was often to be repeated in response to those who, in their eagerness to develop biological reasoning along physical lines, used such analogies freely and uncritically.

Having established to his satisfaction the conceptual basis for a physical interpretation of biological systems, Lotka proceeded to unfold the evolutionary perspective which formed the core of his program. Evolution, however, was not to be understood as the change of species over time, a problem which Lotka had dismissed as of secondary importance as early as 1912, when he drew a distinction between what he called intragroup evolution and intergroup evolution.[40] Evolution in general was taken in a broad sense to refer to any transfer of matter or energy within a system, a definition which, as I shall discuss later, was an extension of the point of view developed by Herbert Spencer in his *First Principles*. *Intra*group evolution was concerned with changes in the distribution of matter within a kindred group of organisms, or, in other words, with transfers of genetic material and the change in the character of a species over time. *Inter*group evolution referred to changes in the distribution of matter between several component groups of a system, such as that occurring among the organisms along a food chain. Intragroup evolution, or change of species, was viewed as a special case of intergroup evolution. Only the intergroup evolution afforded the opportunity to regard the system as a unified whole, which was for Lotka the more interesting approach. As he boldly asserted in the *Elements*:

> Biologists have rather been in the habit of reflecting upon the evolution of individual species. This point of view does not bear the promise of success, if

> our aim is to find expression for the fundamental law of evolution. We shall probably fare better if we constantly recall that the physical object before us is an undivided system, that the divisions we make therein are more or less arbitrary importations, psychological rather than physical, and as such, are likely to introduce complications into the expression of natural laws operating upon the system as a whole.[41]

Lotka was aiming at a law of evolution that would function in the same way, with the same degree of generality, as the laws of thermodynamics, in particular the second law of thermodynamics. This law stated that in an isolated system, only those processes could occur in which the entropy of the system (the quantity of unavailable energy) either stayed the same or increased. The law established a single direction for the processes occurring in an isolated system and could therefore be considered as a law of evolution. Lotka's understanding of evolution as the "history of a system undergoing irreversible changes"[42] echoed a discussion of the second law of thermodynamics as a law of evolution put forth by physicist Jean Perrin in 1903.[43] Perrin himself felt that his formulation was in keeping with the meaning implied by the term "entropy," which had been derived from Greek roots signifying transformation or evolution.

The evolutionary law of the physicist, the law of entropy increase, merely suggested how a law of organic evolution might be stated in terms of energy transformation: it was not itself sufficient to serve as that law. First of all, the second law of thermodynamics was expressed in reference to isolated systems, whereas the organic system received a continual supply of energy from the sun. This meant that the final stages toward which the two systems were evolving were fundamentally different. The isolated physical system evolved toward a true equilibrium, whose end point was determined when the entropy, the amount of unavailable energy, was at a maximum. The organic system, however, because of its continual supply of energy, evolved not toward a true equilibrium but toward a stationary state, where entropy was not at a maximum. This state could not therefore be predicted by a maximum principle like the second law of thermodynamics.[44]

Second, the laws of thermodynamics derived their usefulness from their complete generality: they indicated the direction of energy transformation without regard to the actual mechanisms involved in the changes. In the organic system it was not possible to ignore these mechanisms because they might well influence the course of events from an energetics standpoint. Any law of organic evolution had to encompass the mechanisms by which energy was accumulated and distributed in the world.[45]

But there already existed a principle that took those mechanisms into account: the principle of natural selection, or as Lotka preferred to call it, the "principle of the persistence of stable forms."[46] Lotka's "law" of evolution was a restatement of the law of natural selection, expanded to answer the larger question: to what end did natural selection lead with respect to the energy flow of the organic system taken as a whole?

Ludwig Boltzmann had pointed out as early as 1886 that organisms were engaged in a struggle for energy.[47] Lotka considered in more detail how this struggle might occur.[48] In the first place, those species that were most efficient in obtaining energy for their own use would gain an advantage in competition with other species. The advantage would show in the relatively greater numbers and mass of these organisms. On the whole, such competition for energy would result in an increase in the total amount of organic matter in the whole system, as each organism strove to grow and to reproduce to its maximum capability.

But no matter how much energy was captured by the whole organic system, there would always be more energy reaching the system than could be utilized. Given this excess of energy, a species might arise that possessed superior energy-capturing devices, which would enable it to use that portion of the available energy that was not already tapped by the whole system. Lotka felt that whenever an organism would arise that was able to use energy not already embodied in the organic system, natural selection would work to preserve it.[49] Here again, the result would be an increase in the total mass, or the total captured energy, of the system as a whole.

To the modern reader, this argument sounds as though Lotka was thinking that species evolved structures which enabled them to capture more energy, for example, a plant evolving a special leaf structure. Actually, Lotka was not thinking along these lines: rather he was thinking more specifically of the human species, whose technological capabilities had made it possible to use sources of energy, such as fossil fuels, which were unavailable to the rest of the system.[50] Though he was not always explicit, this concern with society and technology was an important recurring theme in the *Elements* and in his earlier work, and is crucial to the understanding of Lotka's perception of evolution.

These activities were all directed toward increasing the mass of the system, but evolution might also tend to increase the rate of circulation of matter through the system. A faster rate of circulation would represent a more efficient use of energy. Once again, Lotka was thinking in terms of the human species. If a way were found, for example, to grow two crops per year on a given piece of land instead of one, then the same land might be able to support twice the population. The result would be to increase the rate of circulation of matter through this part of the system by speeding up the growth cycle and the period of turnover. An organism that was able to speed up the rate of circulation of matter should, Lotka felt, also be preserved by natural selection, because the result of the faster rate would be increased population.[51]

With regard to the system as a whole, therefore, natural selection would have two effects: it would increase the total mass of the organic system, and it would increase the rate of circulation of matter through the system. Since any transformation of matter was always an energy transformation, the result would be to increase the total energy flow through the system, so long as both energy and raw materials were present in excess. All of these effects would naturally be limited by the constraints imposed by the system itself; the constraints being the environmental conditions and the laws of heredity and variation, which would determine just what forms could arise at a given time.

From this argument, Lotka derived his law of evolution: "Evolution proceeds in such direction as to make the total energy flux through the system a maximum compatible with the constraints."[52] The image he drew upon was that of the world as a gigantic overshot mill wheel, receiving a stream of energy from the sun and discharging it as heat, in a manner analogous to the water flowing over the wheel. Natural selection, in keeping with this analogy, would operate either to enlarge the wheel (increase the mass of the system), or to cause it to spin faster (increase the rate of circulation of matter). The result in either case would be to increase the flow of matter, and therefore of energy, through the system. To the extent that man was participating in these two processes, both enlarging the wheel and causing it to spin faster, he was also unconsciously fulfilling this law of nature.[53]

Lotka's convoluted discussion was more than a complicated way of stating that natural selection was the law of evolution. His intention was to give the principle of natural selection greater generality, so that the problem of evolution could be reinterpreted in light of the diverse mass and energy relationships that cut across species boundaries. He was still interested in intergroup rather than intragroup evolution. What he had done in effect was to make natural selection into a fourth law of thermodynamics.[54]

Wilhelm Ostwald had made a similar but more vague argument in his book *Natural Philosophy*, published in English translation in 1910, where he had also used the comparison between the mechanism of life and a waterwheel.[55] This analogy would have been familiar to a physical chemist: the idea of the mill wheel was the starting point for Sadi Carnot's analysis of the ideal heat engine, which was later reformulated by Rudolf J. E. Clausius into the second law of thermodynamics.[56] Ostwald went on to suggest that organisms which could transform energy most efficiently would be the more perfect, adding that this perspective would be especially important in evaluating the progress of human civilization.

Apart from the explicit origin of Lotka's ideas in the energetics of Wilhelm Ostwald, his attempts to construct a program of physical biology around a holistic interpretation of the law of evolution owed a great deal to the inspiration of Herbert Spencer. Lotka's indebtedness to Spencer is most visible in his early work, where references to passages from Spencer form the starting points for the arguments developed in those articles. In particular, Lotka modeled his definition of evolution on the one Spencer had deduced in his *First Principles*. This definition, it may be recalled, described evolution as the integration of matter and dissipation of motion, with matter progressing from a homogeneous to a heterogeneous state, and motion undergoing a parallel transformation. Lotka formulated his own version in 1911 in an article appearing in Ostwald's journal *Annalen der Naturphilosophie* (later reprinted in English in *Scientific American Supplement*, where he was working as an editor). He was trying in this paper to give a quantitative interpretation of organic evolution, in view of the fact that evolution was attended by physical changes which were in principle quantifiable. The article began with a general statement about evolution, which Lotka considered to be in accord with Spencer's definition:

> The evolution of a given material system is a process which may be expressed as the progressive change in the distribution of matter among specified components of the said material system, through a series of steps taking place in accordance with the principle of the persistence of stable forms (survival of the fittest).
>
> Every change in the distribution of matter in a given system, under given conditions, is accompanied by a definite energy change. Therefore the laws which govern energy changes are laws governing evolution.[57]

Lotka had added the notion of the survival of the fittest (for his definition here only applied to organic evolution), and he had not followed Spencer's reasoning in relating all processes to the persistence of some abstract "force"; but the two definitions reflected similar ways of looking at the world in terms of fundamental physical processes: transformations of matter and transformations of energy (Spencer's "motion"). The difference lay not so much in Lotka's more modern understanding of energy, but in the respective places which each statement held in the overall programs of the authors. Lotka perceived that Spencer's formula represented not the setting of a problem, but the solution of a problem; that problem being to arrive at a formula expressing the changing relations in the course of evolution. Lotka's definition and the important proposition presented in the last sentence ("the laws which govern energy changes are laws governing evolution") were intended to point the direction for the subsequent analysis of the problem; to examine in as quantitative a manner as possible the physical basis of evolution. What was for Spencer a conclusion, was for Lotka only the beginning.

An additional parallel between Spencer and Lotka appears in their understanding of the relation between the whole and the parts. Both were interested in the evolution of entire systems and in the formulation of general statements that applied to this type of evolution. But both realized that any concept of evolution had to be referable to the transformations taking place in the parts. The changes undergone in the parts were merely aspects of the same evolutionary process occurring in the whole system. As Spencer concluded toward the end of *First Principles*:

> So understood, Evolution becomes not one in principle only, but one in fact. There are not many metamorphoses similarly carried on; but there is a single metamorphosis universally progressing. . . . And this holds true uniformly, regardless of the size of the aggregate, regardless of its inclusion in other aggregates, and regardless of the wider evolution within which its own is comprehended.[58]

Lotka held the same view. Like Spencer, he recognized the need to relate the changes observed in the larger aggregates of species and populations to the activity of individuals, and like Spencer, he recognized that the problem was economic in nature, a problem of determining how individuals ought to allocate their energies in different activities. But Spencer's analysis had been full of broad generalizations supported by a few qualitative examples. Lotka, who believed that the best definition was a quantifiable one, went much further by using the mathematical techniques of economic theory for the examination of issues that Spencer had raised. Where Spencer used economic terms still on the level of metaphor, Lotka was able to construct in rudimentary form what was quite literally an economy of nature.

The Economy of Nature

The search for an exact expression of these economic principles naturally led Lotka to the mathematical school of the nineteenth-century economists represented by Augustin Cournot, Léon Walras, Hermann Heinrich Gossen, and William Stanley Jevons.[59] These were a school not in the sense of following the same program, but in that they independently explored the use of mathematics, and especially of calculus, in economic analysis. All but Jevons were unsuccessful in their attempts to popularize their techniques, and it was to Jevons's *The Theory of Political Economy* that Lotka turned for his principal model, adding a few modifications culled from the early work of Vilfredo Pareto.[60]

Lotka's use of Jevons reflected not so much his agreement with the conclusions of that branch of economic theory, as the fact that the mathematical treatment made it easy to transfer the analysis to a general biological context. Jevons's economics was rooted in the Benthamite "hedonistic principle," which related action to the increase of pleasure and lessening of pain. Viewing economics as analogous to the physical sciences dealing with statics and equilibrium, Jevons tried to develop a program of scientific economics from Bentham's doctrine, creating out of the combination a "calculus of pleasure and pain."[61]

Herbert Spencer, also following the utilitarian tradition, had meanwhile framed the hedonistic principle in a psychological and biological setting. He connected activity first to feelings of pleasure and pain, and ultimately to the idea of fitness. That is, those species in which pleasurable feelings were closely correlated with activities conducive to the support of life would survive the longest and be the most fit.[62] Lotka essentially combined the biological expression of Spencer with the mathematical formulation of Jevons.[63] Reasoning that the fitness of a species depended on the way the individual distributed its labor among various activities, he assumed there would be some particular distribution which would produce an optimum benefit (and therefore greatest adaptation). Such a distribution could only be attained if the individual were capable of valuing things at their "true" or "objective" value. The problem was to discover how to determine the objective value in biological terms, in other words, how to relate value to biological fitness.

Lotka's original forays, published in 1914 and 1915,[64] were closely parallel to the parts of Jevons's discussion dealing with the concept of value and the theory of labor, especially his discussion of how labor might be divided to produce the greatest amount of utility with the least amount of pain. In the biological counterpart, individuals would distribute their labor to make the rate of increase per individual a maximum: in this way, Lotka arrived at a definition of the value of a commodity in relation to the rate of increase, which at the same time was a measure of fitness.[65] He expressed this relation mathematically as:

$$V_j = \frac{\partial r}{\partial m_j}, \tag{2.1}$$

where V_j is the objective value of a given commodity; r is the rate of increase per head (which, assuming exponential growth for the population, is equal to the birthrate minus the deathrate); and m_j is the mass of the given commodity consumed per unit time per head. This equation related the value of a commodity (such as a foodstuff) to its effect, when consumed, upon the rate of increase of the individual consumer, all other variables being held constant.

Once the parameter r (rate of increase per head) had been designated as an index of fitness, it was fairly straightforward to dissect the rate of increase into the various demographic and behavioral components which affected it: these were the individual mechanisms underlying fitness. An appropriate question might be, for example, how *efficient* is a given behavior in relation to fitness, as gauged by r? Or how do *errors* in the valuation process come to influence this efficiency?[66] These questions remained abstract and qualitative, however, for Lotka did not have specific numerical examples of these relationships. Energy was certainly of value to the organism, and the worth of a given commodity in energy terms might well be gauged by its contribution to the rate of increase, but this was a far cry from actually determining the value numerically. There was no known equivalent in the animal community to the standard of measurement represented by market prices in the human community.

In Lotka's more mature treatment in the *Elements*,[67] these problems remained unsolved: he could only conclude that the formulas were useful in the relations they revealed between economic and biological quantities, even if they could not be applied to numerical examples. He also downplayed in his later work the connections between his ideas and those of Spencer and Jevons, adopting a game-theory approach based on an analogy with chess to build up his argument. Likening the relations of the organism to the environment to that of chessmen on a chessboard, he imagined that each organism carried around with it certain "zones" of influence and mobility, according to its specific sensory and motor capabilities. The dimensions of the zones determined how the individual could interact with its environment, just as the rules of chess determined how the chessmen could move across the board.

The interesting questions arose by considering how these relations could be changed. First, one could change the character of the zones (the sensory and motor apparatus): this was equivalent to changing the rules of the game, allowing the chessmen to move differently. Second, one could change the relation between the organisms and the environment, while leaving the zones intact: this was equivalent to changing the strategy of the players. Lotka then asked what effect these changes would have on the rate of increase. Reverting to his earlier economic analysis, he arrived at the principle that, given free choice, the behavior of the individual would favor the growth of the species. However, Lotka recognized that a real organism did not consciously maximize its rate of increase, rather it maximized some other quantity analogous to "pleasure." Accordingly he modified his conclusion to arrive at a definition of a well-adjusted species as one whose behavior was adjusted to maximize pleasure, which would automatically also maximize its rate of increase. This brought him back to Spencer's formulation of the hedonistic principle.[68]

The economic method raised two novel perspectives on ecological relationships. One was the quantitative study of resource allocation and distribution of labor, incorporating the idea of choice or strategy along the lines of the chess-game analogy. The other was the relation of energy consumption and behavior to life history and reproductive strategy. Ecologists were just beginning to think in terms of energy and efficiency, the first numerical calculations of efficiency in plants being made in 1926 by E. N. Transeau. In modern ecology, the study of energy distribution using economic models has been developed into a very fruitful area of research.[69] But while Lotka's economic methods anticipated these later developments, they did not serve as a precursor to them. Though we now recognize the importance of economic thinking as applied to biology, and can admire Lotka's imaginative attempt to develop a literal economy of nature, it is easy to see why his methods carried no weight in biology at the time. His interests were mainly in the human animal: when he dabbled in economic biology, he thought like an economist and not like a biologist. In trying to develop a discipline of physical biology, he was too alienated from a biologist's way of thought. His plight brings to mind Wordsworth's lines,

> A primrose by a river's brim,
> A yellow primrose was to him
> And it was nothing more.

This biological naiveté was apparent in his linkage of economic theory to evolution understood in the holistic sense, rather than to evolution of species, where the idea of reproductive strategy would have made more sense. Lotka's idea that the tendency of individual economic behavior was to maximize reproduction and growth was of interest to him only in that it explained the tendency of the whole system to maximize energy flow in the course of evolution. Of course natural selection was part of this mechanism, but Lotka was not trying to extend the understanding of natural selection, only to show how the behavior of the individual parts (seen as energy transformers) was consistent with the operation of the whole (seen as a giant engine). He was not looking forward to the work of R. A. Fisher, but backward to Herbert Spencer. Spencer had made a similar argument in connecting the antagonism between individuation and genesis to the harmonious functioning of the larger system. Lotka was able to make the comparison more forcefully by the use of equations and the concept of energy, but he was not able to frame his questions in a way that set out a program of biological research which others could follow.

Science and the Body Politic

Lotka's concept of evolution, with its affinity to Herbert Spencer rather than to Charles Darwin, was a reflection of the fact that the motivating vision behind his work was a concern with human society and the impact of technology on human development. He was not merely presenting a way of analyzing biological phenomena, but a reasoned worldview centered on the individual's predicament in a time of overwhelming technological expansion. His interpretation of evolution and his "law" of energy flux are only understandable if we keep in mind that his purpose was to demonstrate the unity of man and nature, to show that human activity was intimately tied in with the operation of the vast world engine. Lotka's book was more about human society than it was about biology.

This interest in man was evident as early as 1907, in one of the first articles in the series that was later incorporated into the *Elements*.[70] The article was intended to show how the analysis of the accumulation of chemical substances in a reaction could be viewed as a special case of a wider problem—the study of the distribution of matter in any type of aggregate, including a human population. Lotka's argument, however, proceeded in the opposite direction. He began with the case of a human population and made a demographic analysis of growth as a function of various population parameters, such as birthrate, deathrate, and life span. He then turned to the case of chemical aggregates and, using the same demographic parameters, analyzed the growth of a chemical substance as though it were a case of population growth. Thus he spoke of the "length of life" and "age" of the molecules, the "number of survivors" and the "struggle for existence," concluding that "chemical action clearly presents itself as a case of 'Inorganic Evolution.'" Lotka's use of demographic methods reflected the fact that demography was one area where mathematical analysis had been developed, but it was also a reflection of a very early and more general interest in human populations.

Related to Lotka's interest in population growth was his lively concern about the impact of technology on society. This concern, though evident in Lotka's early papers,[71] was not presented in any great detail until the *Elements*, where it took the form of a philosophical discussion of the essential unity of man and nature. He had originally planned to write a separate book called *Science and the Body Politic*,[72] which was to be an overview of the various ways in which humans had extended themselves through technology. He never completed this book, but he incorporated many of the ideas that would have belonged there into the *Elements*.

The starting point was his image of technological aids as the sensory and motor organs of the social organism, the body politic:

> Man and machines today together form one working unit, one industrial system. The body politic has its organs of sight and hearing, its motive energies, its moving members, in close copy of the primitive body of man, of which it is a magnificent and intensified version.[73]

This concept of society was related to similar ideas expressed by both Herbert Spencer and Wilhelm Ostwald. Spencer had developed the analogy between the individual organism and the body politic in considerable detail, to the extent that he had likened the nerve fibers of a vertebrate to telegraph wires.[74] Ostwald compared the evolution of machinery to organic evolution and stressed the importance of studying civilization historically from the point of view of technical science.[75] Both authors viewed human society from diverse aspects—philosophical, biological, psychological, cultural, and economic. Lotka adopted a similarly broad perspective in his own work.

Underlying Lotka's notion of the body politic was his desire to show that the evolution of the social organism through technological expansion was part of a natural process which contributed to the individual's essential unity with nature.[76] His philosophical outlook was openly Stoical. He believed that it was important for people to recognize nature's laws and to learn to work in harmony with nature's schemes.[77] Working with nature did not imply a society governed by the selfish attitudes inherent in Spencer's philosophy, nor the hedonistic visions of a brave new world, but an altruistic society of individuals who had risen above selfishness to become, in Lotka's words, collaborators with nature. This was orthogenesis as applied to human society, a gradual evolution toward a harmonious, cooperative, and efficient society of the future.[78]

Lotka's Stoicism was expressed with an unusual degree of fervent idealism. At a more pragmatic level, his ideas were mirrored in many different forms in the literature of the time, where the theme of industrial expansion and society's response to it recurred often. The tremendous expansion of America's industry at the turn of the century stimulated several pronouncements on man's duty to suppress his unruly individualism and to contribute to the growth of the modern, industrial society.[79] This was the underlying theme of Horatio Alger's popular novels: with wit, daring, and luck the heroes might overcome their impoverished beginnings, but these heroes were no rebels; they ended up by quietly settling into their proper place in society.[80] Whether technology was seen to be a beneficial or malevolent force, the resounding question of the day was: how would people respond and adjust to the changes of the new age?

In 1901, for instance, we find Brooks Adams arguing that nature favored organisms that were most efficient in their use of energy. The future society, according to Adams, would have rulers skilled at administering "masses vaster than anything now existing in the world," and laws and institutions that would "take the shape best adapted to the needs of the mighty engines which such men shall control."[81] As H. G. Wells observed in his visit to America at the turn of the century, the country seemed to be at a turning point in its great surge of growth, a change from the "first phase of a mob-like rush of individualistic undertakings into a planned and ordered progress."[82] Lotka had tried to ground this ideal of a planned and ordered progress in the basic laws of nature. Technology, far from alienating humans, was a means by which they could achieve unity with nature through the combined actions of the body politic, on the condition that they could keep up with the progress of technology and suppress any selfish tendencies that interfered with the social organism.[83]

The Reception of the *Elements*

Lotka's first point of reference for his program of physical biology was therefore a particular vision of human society. The fact that it was welded onto an interesting but biologically naive program, setting out the basis for an entirely new discipline in science, guaranteed that many people would find the book puzzling at the very least. With some people, years of continuous thought eventually crystallize into a single, clear insight; with others, the pent-up waters of philosophical imagination, finally released in writing, tumble forth in a torrent of ideas foaming in all directions. So it was with Lotka. The breadth and exuberance of his work brought forth the following description from his publishers:

> One gets first a faint and vague glimmer and eventually a far-flung and inspiring view of the universal evolving system, functioning under the scope of physical law; and one turns to the universe, like a Demiurge for whose pleasure the stupendous drama of the eternal ages is being enacted. High-sounding language, to be sure. But the book calls it forth.[84]

This description did in fact accurately capture the effect that Lotka hoped to achieve. But the practical side to his nature told him that the book's success depended on his ability to convince modern specialists that his work was relevant to their particular needs. Estranged from the biological community, however, and still thinking like a physicist, he failed at first to identify the audience that might have ensured his book's maximum success. Even more importantly, perhaps, he lacked a close collaborator who could translate the book's ideas into the practical problems of the biologist.

Review copies quickly went out to statisticians, physicists, biometricians, and epidemiologists, but none went out to ecologists.[85] Lotka did not think of a review in *Ecology* until Charles Adams wrote to him, and even then he did not take Adams's enthusiasm fully to heart, but continued to worry because physics and chemistry journals were not reviewing the book. Perhaps, he reasoned, they had been put off by what seemed like exaggerated claims of novelty in the advertising brochures.[86] In 1927 he still complained that the main response to his book had come from biologists: "I am convinced that the main efforts remaining to be made to advance the work must come from physicists rather than biologists, for the general principles involved are physical rather than biological; it is the details that are biological."[87] The reviews of the book were in fact generally favorable, with the exception of a mixture of praise and petty complaint in *Science* by Edwin Bidwell Wilson, a statistician at Harvard, who began by placing Lotka's book within his own category of "really new books" that included such works as "'The Fitness of the Environment,' 'Winnie the Pooh,' 'Die Ausdehnungslehre' or 'Oedipus Tyrannus.'"[88] He concluded by stating that it was beyond the ability of any one reviewer to describe the book's merits or to assess its faults. But even of the other reviews, which were much more balanced, Lotka expressed disappointment that they all failed to "hit the spot":[89] the reviewers, he felt, seemed unable to grasp exactly what the book was about.

Seeking to calm his mounting anxieties, the publication sales manager of Williams and Wilkins assured Lotka that his book was selling reasonably well for its kind. From an original edition of about 2,500 it sold 568 copies in 1925, followed by an average of 234 per year until 1930, when sales dropped sharply off to 73.[90] A small market for the book continued throughout the 1930s, until by 1940 the edition was nearly sold out.[91]

During these years, Lotka gradually turned away from physical biology and devoted himself wholeheartedly to mathematical demography, a field in which he had always held an interest and which was an offshoot of his work in physical biology. After completion of the book, he accepted a position immediately as supervisor of mathematical research at the Metropolitan Life Insurance Company in New York. In 1934 he was promoted to the executive rank of assistant statistician and remained there until his retirement from the company in 1947. In demography, Lotka soon attained a high reputation. According to the later assessment of one sociologist, his work in stable population theory was comparable to that of Newton in physics, in that it "achieved a synthesis in analytical theory which had far-reaching significance," and helped to "set a frame for new empirical investigations," some of which ultimately revealed limitations to how the theories could be applied.[92] After joining the establishment, Lotka's extravagant visions of man unified with the universal forces of nature gave way to a pragmatic concern with problems of population growth and their scientific study. With more restrained imagination, Lotka now expressed himself through work in professional population associations, such as the International Union for the Scientific Investigation of Population Problems formed in 1928 by Raymond Pearl (later the International Union for the Scientific Study of Population), and the Population Association of America (formed in 1931 with Henry P. Fairchild as its first president).

The *Elements of Physical Biology* failed to inspire the creation of the new discipline that Lotka had defined, but it did find a small, appreciative audience among ecologists who were searching for ways to organize their science. Charles Adams found much to admire in the systems approach and its emphasis on energy relationships: Lotka's idiosyncratic conception of evolution would not have appeared so odd to the proponent of the dynamic-process viewpoint, for he had already come to adopt a similar perspective in his meditations on the meaning and direction of ecological science. A second equally appreciative audience of ecologists was attracted to Lotka's mathematical treatment of predator-prey relationships, and to his demographic analyses, which held the promise of applications to animal populations as well as human. The way in which these ideas became incorporated into ecology owed a great deal to a coincidence: just as Lotka's book was published, Vito Volterra, the eminent mathematical physicist, published his own account of the mathematical theory of the struggle for existence. Some of Volterra's work overlapped with Lotka's, and as Volterra's results were made known and used, Lotka's reputation also benefited, thanks in part to Lotka's vigilance in making sure that his priority was secured. These theoretical researches coincided with an increased interest in populations within the different branches of the ecological community. I shall begin chapter three with a sketch of some of these new interests before returning to Lotka's discoverer, Raymond Pearl.

3

The Quantity of Life

The Ecology of Populations

To recapitulate, ecology had by the 1920s developed along two paths, each leading to the opposite end of the spectrum ranging from the individual to the community. "Individual ecology" or "autecology," as it was sometimes called, focused on the study of whole organisms and their relations to the environment: this style of ecology was closely connected to physiology. At the other end, "community ecology" or "biocoenology" or "synecology" was the study of the larger natural associations of species, both in space (biogeography) and in time (succession). Connecting these two levels were field and experimental studies of competition and food relations, aimed at revealing how the cumulative interactions between individuals give rise to larger, stable communities. Ecology was mainly a descriptive and experimental science already confronting the problem of organizing information. Some ecologists, acutely aware of the need to give order to the science, were actively searching for principles that would help them to see the constant patterns behind that which was always changing. They still felt very much in the dark.

We now come back to that aspect of nature which particularly fascinated Darwin: the quantity of life, as seen by the great powers of increase of organisms, and the way that quantity is held in check. The question of numbers, of how they vary and what causes them to vary, is of particular concern in economic biology. It was largely because of the importance of these economic problems that ecology began to turn in the 1920s to the science of populations. Economic biology had always been closely tied to the growth of ecology, although it was not identical with ecology, the latter being broader in scope. But the promise of sound resource management was the most important reason for ecology's continued existence. The increasing necessity to address problems in economic biology in the twentieth century was in turn reflected in a new interest in the study of populations among ecologists.

Economic entomology, for instance, was fast gaining in importance following improvements in storage and refrigeration techniques, which allowed more goods to be moved over larger distances and increased the chances of importing harmful insects. The Argentine ant, Mediterranean fruit fly, European corn borer, and Japanese beetle, among others, were added to the list of serious imported pests in the early 1900s. Their cumulative effect was to stimulate the growth of economic entomology and to attract attention to the idea of biological control as an inexpensive and effective adjunct to the established methods of chemical control.[1] Biological control, which referred to the control of pests through the use of beneficial parasites and predators, had achieved spectacular success in 1888–1889 with the introduction of the ladybird beetle into California to control the cottony-cushion scale then threatening the young citrus industry.[2] Within a season the beetle had effectively reduced the pest and saved the industry. Since that time, the U.S. Bureau of Entomology, organized in 1904 from the Division of Entomology of the U.S. Department of Agriculture, had also turned to the study of biological control: it soon attained prominence in the field, although its work did not match the success of the ladybird in California. In particular, under the direction of Leland Ossian Howard, the bureau carried out experimental projects on the parasites and predators of the gypsy moth, brown-tail moth, corn borer, and Japanese beetle.

The First World War further intensified the work of entomologists, especially in the areas of grain storage, lice control, and crop production.[3] Throughout 1917 and 1918 the Bureau of Entomology made its first attempt to compile a census of insect damage and to systematically record the increase of crop pests all over the country, information which was then distributed to interested parties from Washington, D.C. At the same time, Howard noted that the war had actually hampered insect control measures by disrupting the supply of arsenical compounds used widely in insecticides. This experience may well have impressed upon him the value of exploring other methods of control more intensively.

The more biologists began to study these problems, the more they awakened to the full extent of their ignorance. The dramatic appearance of insect epidemics common to the experience of entomologists reinforced their awareness of how little they knew about nature's laws and how far applied science stood from the ideal of total control. The following is one entomologist's description of a few of the insect outbreaks of the 1920s and 1930s:

> The appearance of spruce budworm moths when in flight during an epidemic has been compared to a snowstorm in the tree tops. The forest tent caterpillars, when moving over the ground as they do, in search for undefoliated trees, have been so numerous that their crushed bodies on railroad tracks have made it impossible to pull trains up even slight grades. More recently this insect is said to have interfered similarly with automobile traffic

> on cement highways in infested areas. . . . Walking sticks become so abundant that their eggs falling on the ground suggest the patter of raindrops. This is truly no exaggeration.[4]

As these outbreaks were studied in greater detail, it became clear that population events reflected a more complex fabric of causes than had previously been imagined. The early success of the ladybird in California stood revealed as an exceptional case which did not warrant the mood of popular faith in biological control that had followed it.[5] Biological control remained an important idea, but entomologists were less confident in its ability to achieve results that were economically significant. As Howard expressed it, summarizing the control work of the Bureau of Entomology up to 1930: "I do not waver in my unfailing belief in the basic value of the principle of biological control, but my outlook becomes more or less confused when I consider the implications."[6]

The emphasis given to biological control by economic entomologists did, however, lead to new insights about how various factors were acting on population numbers. In general these factors were seen to act in three ways.[7] Some served to restrict the multiplication of species at those times when conditions were most favorable for increase: these were the biotic agents, such as parasites and certain predators, whose rate of increase was directly tied to the rate of increase of their hosts or prey. Other factors could cause a constant percentage of destruction no matter what the densities were of the populations destroyed: these were the climatic agents, such as drought or frost. A third category was composed of predators, such as insect-eating birds, whose populations were not correlated with the fluctuations of any one prey species. Because of their varied diet, the action of these predators was not density-dependent, but neither was it comparable to the effects of climatic factors because the percentage of mortality was not necessarily constant. In fact, such predators might destroy a higher percentage of prey when the prey were at low population densities rather than at high densities.

The third category was not thought to be important in the control of populations, so attention was directed mainly toward the first two, which were labeled under various headings by different biologists. L. O. Howard and W. F. Fiske, for example, distinguished between "facultative" and "catastrophic" factors, respectively, as early as 1911. Their colleague W. R. Thompson referred to "individualized" and "general" factors with the same idea. The terms "density-dependent" and "density-independent" later came to be the preferred designations and are still in use. By the 1920s, therefore, economic entomologists had begun to describe and to classify the assorted checks to population increase, but there was little agreement yet about how an individual factor could best be classified and what its relative importance was in regulating populations.

In England, Charles S. Elton drew attention to the need for ecologists to study population fluctuations in mammals as well. Elton's population interests were a diversion from his main interest in community ecology, to which he eventually returned; but they were an important diversion, for, as he pointed out, "the study of the regulation of animal numbers forms about half the subject of ecology, although it has hitherto been almost untouched."[8] During the 1920s Elton had been occupied with an analysis of small mammal cycles, using data from the Hudson's Bay Company in Canada. In 1924 he summarized what was known about the causes and effects of the changes in the numbers of various populations, drawing on his own research and on the observations others had made on periodic epidemics.[9] He tried to show that these fluctuations, especially in small mammals such as lemmings and hares, could be correlated with climatic variations, the most important being the eleven-year sunspot cycle. Knowledge of these correlations was of considerable economic interest in the fur trade, but Elton's reasons for studying them went beyond economic applications. The ecological study of populations, he felt, was also important for the insights it might give to a number of puzzling problems in evolutionary theory.

Elton could hardly have avoided developing a sense of the evolutionary implications behind ecological questions. As a student and then as a professional biologist at Oxford, he was steeped in the neo-Darwinian tradition passed from E. Ray Lankester, through E. S. Goodrich, to the generation of J. B. S. Haldane and Julian Huxley. Elton was himself a student of Huxley's, and he clearly absorbed some of his teacher's characteristic ability to see some bearing on evolutionary issues in a variety of facts culled from diverse disciplines. These issues and their relation to population studies were discussed in Elton's article of 1924 and then in his influential little book of 1927, *Animal Ecology*.[10]

Elton felt his treatment of evolution, though brief and speculative in tone, to be a departure from most ecological writings of the twentieth century, which were focused on problems relevant to geographical distribution and ecological succession, on the one hand, and physiological functions, on the other. These questions could be broadly considered to be within the domain of evolutionary biology, in the sense that they were intended to unravel the mechanisms of the struggle for existence. But ecological writings on the whole did not directly address problems in evolution or questions about the role of natural selection. Ecologists had become caught up with the day-to-day processes in nature; ecology had moved away from a specific concern with evolutionary puzzles.[11] Elton perceived this separation between ecology and evolution to be so great that he began his chapter on evolution with half an apology: "It may at first sight seem out of place to devote one chapter of a book on ecology to the subject of evolution."[12] But ecology *was* relevant to evolutionary theory if the study of populations was taken to be an integral part of the discipline, and this was the point that he wished to make.

He singled out two evolutionary problems which had long been subjects of unsettled debate involving the role of natural selection in producing adaptations. One problem was animal coloration: concealing color patterns were obviously adaptive in a general way, but why one particular pattern and not another? How could both color forms in dimorphic species be adaptive? The second problem was the origin of differences between closely allied species: large differences between genera were clearly adaptive in many cases, but the often minute differences between closely allied species appeared to be trivial and meaningless. R. A. Fisher's contention, expressed in *The Genetical Theory of Natural Selection*, that organisms were "marvellously and intricately adapted, both in their internal mechanisms, and in their relations to external nature,"[13] was by no means accepted by systematists, to whom the role of natural selection seemed decidedly limited.[14]

Elton's speculations, occupying a middle ground between the rejection of natural selection and the wholehearted acceptance of its power to produce specific differences, fell into line with the current systematist's viewpoint.[15] He argued that in populations which underwent periodic and drastic reductions in numbers, natural selection must necessarily operate on different characteristics at different times. When conditions were crowded, for instance, selection might favor resistance to disease or ability to escape enemies. During minimum population years, it might act instead on the ability to resist climatic conditions, or it might temporarily cease, allowing neutral traits to become lodged in a population by mutation. The result of all these different selective forces, acting with varying strengths as the populations fluctuated, was that characters would be alternately acquired, weeded out, or simply kept as indifferent: these indifferent traits, acquired at a time when selection was not acting, that is, when the struggle for existence had ceased, could account for the existence of apparently nonadaptive differences between closely allied species or of particular color patterns. Natural selection need not be invoked as an explanation for every minute characteristic, or as Elton expressed it, "There is, so to speak, an entrance examination by natural selection, which weeds out the worst candidates, but the final examination is by lot."[16] Elton fully admitted that these explanations were not quite satisfactory and left much room for disagreement.[17] His intention, however, was not to explain the intricate nature of adaptation, but rather to show that ecological studies in relation to animal numbers could be used to illuminate more general evolutionary problems.

At the same time that Elton was writing about populations and evolution, a new interest in population studies also arose in physiological ecology. These studies emphasized laboratory experiments and were largely attempts to reduce nature's complexity to manageable proportions by bringing a part of nature into the laboratory. As in the field studies of the other ecologists, the first efforts to understand what was going on led mostly to a heightened awareness of ignorance. At Yale University, Lorande Loss Woodruff experimented on succession in laboratory populations of protozoa: these experiments left him feeling more bewildered than informed. After unsuccessfully trying to dissect the sequence of events in a hay infusion microcosm, he ruefully concluded:

> The competition between the various forms is so keen and the cycle is so rapid that even daily observations are, at times, insufficient to reveal the kaleidoscopic changes. . . . One who follows a series of infusions day by day cannot but be impressed with the intense struggle for food and the eternal warfare in this microcosm, and become convinced, though he cannot prove, that in the final analysis the paramount factor is food, though many other factors, such as excretion products, etc., may play a not unimportant part.[18]

In Chicago, on the other hand, Warder Clyde Allee would soon extend physiological ecology with more success to the study of single-species animal aggregations.[19] He realized that it was essential to study aggregations, not only individuals, from the physiological point of view, because other members of a population could and did influence an individual's ability to tolerate stressful environmental conditions. Allee was especially interested in the origins of proto-cooperative behavioral responses in animals. This was ecology in its original sense, "scientific natural history," or the study of the relations between organisms and the environment. But the title of Allee's book, *Animal Aggregations: A Study in General Sociology*, indicates the particularly close ties between ecology and sociology that was the trademark of the Chicago school and that undoubtedly fostered an interest in populations.

Economic entomologists were also turning to the laboratory. In the late 1910s, Royal Norton Chapman at the University of Minnesota had begun to study the biology of the flour beetle, *Tribolium confusum*, which was a widespread cereal pest in the state.[20] Attention had been drawn to this insect during the war, when food was being conserved and several substitutes for flour were in use. Chapman started studying the insects infesting wheat flour and its substitutes, continued his work on the beetle after the war, and by the mid-twenties was actively promoting the laboratory study of populations in ecology.

These population studies were still few and far between: population ecology as such would not be recognized as a distinct branch of ecology, with its own theories and methods, for another twenty years. But they were the start of a new direction in ecological research that would raise a host of new problems about how populations should be studied. The questions involved numbers, and the answers seemed to lie in the direction of mathematics. Before too long the mathematical answers began to appear. They arrived in diverse dress, ranging from straightforward algebraic descriptions of known populations, to intricate and unrealistic models of the interactions of species in an imaginary community. These trickled and then rushed upon the ecological community in the 1920s and 1930s, creating an embarrassment of theoretical riches in a science which had hitherto been staunchly empirical. Some of the ideas were absorbed into ecology over a period of time in a sporadic fashion, the vast majority of ideas being completely ignored. The few ideas which, in the form of simple mathematical models, did enter ecology and formed the kernel of the intellectual tradition that grew into population ecology, often owed their survival to the fact that they were intensely promoted and consequently highly visible. This is where we find Raymond Pearl poised to enter the scene; a self-conscious crusader of scientific rationalism, clutching in one hand a milk bottle full of fruit flies and brandishing in the other his infamous logistic curve.

The Biology of Groups

Raymond Pearl was part of that group of young biologists at the turn of the century who were captivated by the experimental approach, which was providing such a refreshing contrast to the static morphology, directed at animals "thoroughly pickled," of an earlier generation.[21] He was introduced to the new biology by Herbert Spencer Jennings while still an undergraduate at Dartmouth College.[22] Jennings took Pearl with him to the University of Michigan, where he completed a doctoral thesis on the behavior of flatworms. After three years as an instructor in zoology at the University of Michigan, Pearl traveled to Europe in 1905–1906, working at Leipzig, at the Marine Biological Station at Naples, and at the University of London. It was the stay in London that was to prove most crucial for his development, for there he met Karl Pearson and became a convert to the statistical view of nature which Pearson advocated.

Pearl had read Pearson's book, *The Grammar of Science*, in 1900; it made a profound impression on him. Twenty years later he still recommended it warmly to students, even while acknowledging that other biologists would judge its positivistic approach to science not very successful when applied to many areas of biological research.[23] *The Grammar of Science*, first published in 1892, was an attempt to clarify how a scientist builds up a picture of the external world from the raw material of sense perception. Science for Pearson was the orderly classification of facts, followed by the recognition of their relationships and recurring sequences. Once the facts were classified, a scientist had to apply his creative imagina-

RAYMOND PEARL, 1879–1940
Photograph by Greystone Studios, Inc.; from the Alan Mason Chesney Medical Archives of The Johns Hopkins Medical Institutions

tion to express those facts in the most economical way. This process led to the enunciation of scientific laws: "The single statement, the brief formula, the few words of which replace in our minds a wide range of relationships between isolated phenomena, is what we term a scientific *law*."[24]

Formulating the law was only the first stage of science; the law then had to be tested against more facts to make sure it was correct. Furthermore, not all laws were equally interesting or important. The first statement of a law might be only approximate, to be followed by a more general statement covering a wider range of phenomena. By successive approximations, scientists approached the more fundamental laws of nature. A scientific law was therefore similar to a hypothesis; it was a hypothesis which, through further testing, had been shown to accord tolerably well with the facts. Pearson considered his method to be Baconian, although he felt that Bacon himself had not made the method work because the classification of facts had still been in a primitive stage. His model of a successful "Baconian" scientist was Newton; the method of successive approximations of laws was exemplified in the successive descriptions of astronomical motions. Darwin too was "Baconian," according to Pearson, because he was able to classify the facts of biology in the formula of evolution by natural selection.

The value of the scientific method was not restricted to science per se, but was seen to form the basis for all human conduct and planning. Science, in the enlightened age of the future that Pearson projected, would be applied to every aspect of life. He confidently concluded that once the scientific method was diffused throughout society, it would produce a society of individuals free from bias: sound citizenship and stability would be the result. As Spencer had suggested earlier, all knowledge would be unified by a consistent, comprehensive, scientific outlook. Moreover, the scientific method was the only one leading to knowledge, the alternative being to remain ignorant. But science meant progress in the long run. Pearson quoted with approval a remark by W. K. Clifford to the effect that science was not merely a condition of human progress; it was human progress itself. This was the philosophy of science, with its vision of the scientist as the leader in the enlightened future, which Pearl imbibed as a youth and which, like a good disciple, he applied resolutely in the shaping of his own worldview.

His year with Pearson gave Pearl more than a philosophy; it gave him an interest in mathematical approaches to biology.[25] The mathematical development of evolution followed from Pearson's eugenic interests and his belief that the scientific method could bring social evolution under control. In order to make any advances in evolutionary biology, vague terms such as selection or heredity had first to be redefined in precise mathematical form. Following the example of his mentor, Francis Galton,

Pearson tried to establish such quantitative definitions of central evolutionary concepts, and to show that the quantities so defined were the true causes of change. Having arrived at quantitative measures of the factors of evolutionary change, it would be possible to compare the action of one factor against another and to draw sound conclusions about social evolution. For human populations alone, the information was available to allow for this degree of mathematization: this was the body of statistics, accumulated over the previous decades, on fertility, growth, disease, and mortality. Evolutionary biology would have to concern itself with vital statistics if any conclusions were to be drawn about the effects of different selective pressures on society.

Pearl was completely converted by Pearson's arguments. He saw himself not merely as a disciple of Pearson, but as the modern descendant of the mathematical tradition in evolutionary biology extending through Pearson back to Francis Galton. It was this tradition that he was determined to introduce to American biologists.

Upon returning to America, his research interests turned to genetics. Unlike Pearson, Pearl was soon convinced of the truth of Mendelian arguments in genetics. The result was a falling-out with Pearson in 1910 over a Mendelian interpretation of the inheritance of fecundity;[26] an argument which did not affect his regard for Pearson's statistical approach, however. He began research on Wilhelm Johannsen's "pure-line" theory of evolution, a problem which brought him into line with H. S. Jennings's research interests. Both Jennings and Pearl, working independently on protozoa and fowl, respectively, believed they had experimental confirmation of Johannsen's conclusions that fluctuating or continuous variations were unimportant in evolution, thereby strengthening the argument for Mendelism and discontinuous evolution.[27] Jennings eventually settled at the Department of Zoology at the Johns Hopkins University, while Pearl sharpened his statistical skills at the Maine Agricultural Experiment Station, where he stayed from 1907 until 1918. In 1915 he published *Modes of Research in Genetics*, a critical study of current methods in genetics; by 1916 his reputation was solid enough to earn his election to the National Academy of Sciences.

In 1918 Pearl was recruited by William Henry Welch to become the first Professor of Biometry and Vital Statistics in the new School of Hygiene and Public Health at the Johns Hopkins University. Welch's choice was understandable, for Pearl brought an enthusiastic interest to a broad range of problems in statistical biology. He believed that vital statistics, an area unjustly ignored by biologists, opened up an exciting field of biological investigation. He would prove to be a good propagandist for the development of biometry at the new school.

Pearl's initial interests at Johns Hopkins were in the areas of public health and epidemiology, but an accident in 1919 caused him to delve into a different but related problem, that of human population growth. He was nearing the completion of a study on the relation of tuberculosis to environmental factors when, on 27 November 1919, a fire in the building in which his laboratory was housed destroyed the complete work and the raw material on which it was based.[28] Also destroyed were his entire private collection of reprints and pamphlets, and unpublished records for the previous twenty years, including his records of his genetics work carried out in Maine. As if to compound his misery, Pearl had moved into the building just three weeks before the fire and the insurance had not yet been transferred to the new location. After this catastrophic loss, he did not have the heart to redo the tuberculosis study, and he turned instead to an issue that was partly inspired by his wartime work on Herbert Hoover's Food Administration Program, the problem of overpopulation and food supply.

His studies of human population growth were begun in conjunction with a series of laboratory studies on life duration in animals, which carried over from his earlier interest in genetics. Pearl's original intention had been to use mice for the experimental work, but the mouse colony he was building up had been destroyed in the fire as well. On the advice of T. H. Morgan and Jacques Loeb he switched to *Drosophila*, on which there was already a large literature. With four strains from Morgan's laboratory, he began breeding the flies in December 1919.

Pearl believed that the research advanced by Morgan and his group had by no means exhausted the important lines of investigation in genetics. In particular, most people had not recognized the value of vital statistics in evolutionary biology, a neglect which, he could not refrain from commenting, was "another sad example of the slight influence of logic upon human behavior."[29] Adopting Pearson's mathematical point of view and combining it with the experimental style he had learned under Jennings, Pearl set out to build his own program in genetics research. It was essentially a program in comparative demography, the start of a systematic attempt to apply demographic techniques to animal populations. Pearl borrowed freely from the tools of the actuary: he constructed life tables for his flies and calculated deathrates and life expectancies, just as the statistician would for a human population. He showed that there were a few basic types of survivorship curves which organisms having different life histories might be expected to follow. Over the next decade he added gradually to the quantitative data on duration of life in *Drosophila*, while his associates extended this demographic analysis to other organisms. In all this laborious research, Pearl was not simply promoting statistics in biology: he wanted to show that biostatistics was the proper method of a new science entire of itself; that it was "the sign, the symbol, and indeed in some respects the very essence of a *selbständige Wissenschaft*, namely the *biology of groups*."[30]

This "biology of groups" was another version of the several forms of holistic thinking prevalent in the 1920s. Pearl believed it was valuable to focus one's attention on some characteristic of the whole, such as the duration of life of the individual organism. Duration of life was not a biologically separate characteristic, but the "expression of the total functional-structural organization or pattern of the individual."[31] To study the duration of life was to measure, in a single quantifiable expression, the total vitality of the animal. It was this opportunity for quantification that was the chief advantage of the method. In the case of the population, the holistic perspective meant studying such features of the population as growth rate, mortality rate, or fertility rate. Pearl did not go so far as to claim that the population was a supraorganism, for of course it lacked the organism's organization, but he tended to treat his populations as wholes for the sake of scientific argument. This holistic point of view, which Pearl shared with his friends L. J. Henderson and William Morton Wheeler, was to create some confusion later on when critics who did not view populations as aggregate wholes began to criticize Pearl's conclusions.

Whether looking at the individual or the population, the object of the inquiry was always the same: to determine what part of the character of the whole was inherited and what part was affected by the environment. The theme of heredity versus environment ran throughout all of Pearl's work. He set about answering this question by measuring the character under study and then correlating that measure with a variety of other factors. Science became a search for quantitative measures and correlations under different experimental conditions. How did the duration of life vary under different environments or in different mutant strains? How did the shape of the survivorship curve change in different populations? What was the law of population growth? These were the problems he set himself. Though seemingly mundane in origin, they were to prove surprisingly controversial in the hands of the crusading Pearl.

With his arrival at Johns Hopkins, Pearl hoped to find himself in an environment that would allow him to develop his program in his own way. He had barely settled into the School of Hygiene, however, when he began to contemplate the opportunities of commercial science. In April 1923 he was approached by the Fleischmann Yeast Company with a proposal to move to New York as the director of a new research laboratory being planned there.[32] The offer was attractive: he was to have the freedom to do whatever research he wanted, as long as it pertained to yeast. After a month spent toying with the idea, he decided to stay at the university, a choice made easy by a new arrangement the university held out to Pearl.

The chief point of the new agreement was an increase in freedom. As

Pearl wrote to Lawrence Henderson, "I am, so to speak, given the keys of the School of Hygiene and Medical School to do just as I please."[33] Specifically, he was appointed Professor of Biology in the Medical School and was relieved of all routine teaching, apart from one course of lectures each year. Although he was by many accounts an inspiring teacher, he disliked classroom teaching and preferred the closer relationship of master and apprentice; he believed that real learning was achieved only by doing.

By the next spring Pearl had conceived even grander prospects. In March 1924 he set out plans for a long-term project on the duration of life, combining as usual the statistical and experimental methods.[34] He proposed to study the impact of various factors on the duration of life—factors such as population density, starvation, temperature, and heredity. The main subjects were still fruit flies, but he made it a strong selling point of the proposal that he hoped to learn about the factors determining longevity in man. He also intended to make a statistical study of the inheritance of life duration in man, based on family history data, and to study disease and senescence, both experimentally in animals and by analyzing human autopsy records from the Johns Hopkins Hospital. For this ambitious program in general and human biology he wanted no less than an endowment of $1 million to ensure continuous support for the laboratory.

As a possible source of funding, Pearl sounded out the Fleischmann Yeast Company again. But Fleischmann was not willing to put up the funds for such a project at Johns Hopkins where Pearl wanted to stay. He then turned to the Rockefeller Foundation, where he could not have found a better ally than Edwin R. Embree, who shared all of Pearl's enthusiasm for this line of research, especially as it involved eugenics and human genetics. Embree saw Pearl's work as part of a major new program in human biology to be started and funded through the foundation. This program would include a variety of projects, such as primate behavior studies, anthropological research, and research on mental hygiene.[35]

Pearl's request for $1 million was exorbitant, however. Embree suggested a more modest five-year grant of $175,000 instead. Pearl was chagrined. This amount was short of his own operating estimate by $5,000 per year. He accepted the figure with the condition that he be allowed to ask for additional funds if he needed them later. With Embree's support and enthusiasm now ensured, the proposal passed at the foundation in May 1925.[36] Pearl's project, in the shape of a new Institute for Biological Research, was launched, and Pearl prepared to savor his new independence.

As director of the institute, Pearl had finally found the freedom he had long sought. As he wrote to his friend Major Greenwood, the British epidemiologist, "All these people have sufficient confidence in me to set me up in a perfectly free and untrammeled show to do what I like."[37] The institute was an autonomous unit within the university; at $15,000 a year Pearl was one of the highest paid professors at Johns Hopkins; and he was completely free of the disagreeable chore of teaching. Although he did supervise a few doctoral candidates over the next few years, he did nothing special to encourage students. He did, however, offer encouragement to research associates to come to his laboratory and to participate in the institute's program of research. His teaching responsibilities were turned entirely over to Lowell J. Reed, his colleague and occasional collaborator, who enjoyed teaching. As Pearl explained to Greenwood:

> The word "biological" in the title means really freedom—freedom to work on any kind of problem in any kind of way that I want to. For the past twenty years I have been engaged in one way or another in some form of applied science in which overtly or tacitly it was expected that at regular intervals I would put salt on the little bird's tail. From now on, if fortune smiles upon the enterprise, this form of bird snaring is done with for good and all.[38]

On 6 July 1925 he moved into his new quarters in the Medical School, a "snug little shop without any frills or ornamentation," ready to begin his show.

Although associated with the School of Hygiene and the Medical School, the institute was separate enough to ensure Pearl complete control over its activities. The staff hovered around twenty, composed of a core of about five researchers, with a number of assistants and consultants, plus a handful of graduate students and visiting researchers. The small size enabled Pearl to manage the program to a "single concentrated purpose" which, as he readily acknowledged, centered on his own interests. Projects fell into two categories. Under the heading of "human biology" were statistical studies of health, longevity, population growth, and human genetics. The second heading of "general biology" encompassed the same broad range of topics, but consisted of experimental studies on lower organisms. One of the objectives was to show how an experimental attack on such problems might shed light on human biology. As part of the institute's activities, Pearl also founded two journals: *The Quarterly Review of Biology* in 1926 and *Human Biology* in 1929.[39]

Pearl was one of those energetic and restless men who, though they are consumed by but a single intellectual problem, nevertheless express that interest in a flurry of different activities. Pearl was expansive in every respect, being of unusual height and weight and having a broad and lively intellect. With characteristic energy he pursued his course through life as though his methods, and his conclusions, were beacons of rationality in a dark, sometimes illogical world. Characteristic also was his optimism that his pursuits would yield results of fundamental importance, perhaps even a new understanding of life itself. Such men stimulate intense responses in others. Their exuberance may inspire great admiration and excitement, but if they fall into error, they may inspire an equally enthusiastic enmity. Pearl did both. Though he was at the height of his career in 1925, already there were faint rumbles of trouble to come. One of the most controversial issues in Pearl's career was the analysis of population growth to which he had turned in 1920. This controversy centered on his use of a single equation, called the logistic curve, to describe the growth of a population. From this small beginning grew an intensely polemical dispute which lasted well over a decade, ranged widely over the disciplines of biology, demography, and economics, and eventually contributed to population ecology one of its simplest mathematical models.

The Logistic Hypothesis

The First World War had served to reinforce two beliefs in the minds of those who were attentive to the problems of population growth. The first was that wars in general were the direct or indirect results of the pressure of population upon the means of subsistence. The second was that any effective response to wartime stress demanded prompt access to statistical information on those populations concerned. Pearl, who from 1917 to 1919 was Chief of the Statistical Division of Herbert Hoover's Food Administration Program, was one of those who emerged from the war with both tenets firmly in mind.[40] His work had brought home the realization of how close Europe had come to the brink of famine as a result of food shortages caused by the war.[41] But if the dangers of population growth could be clearly felt, it also seemed that predictions of long-term trends were based as much on guesswork as on what might be called a scientific method. Following a brief study of the impact of the war on the birthrates and deathrates of European countries,[42] he began to study the population problem more seriously.

In 1920, working with his colleague Lowell J. Reed, Pearl published the first of a long series of controversial articles on the rate of population growth.[43] Reed, a mathematician, supplied the statistical analyses behind this and subsequent papers, but the topic itself and the conclusions drawn from it more properly belonged to Pearl. In this article they discussed the ways in which growth, either of individuals or of populations, could be represented mathematically. What they sought was a *law* of population growth, by which they meant an equation which would both conform to certain assumptions about how populations must behave and indicate future trends reasonably accurately. Pearl's insistence that the equation he and Reed presented in 1920, later known as the logistic curve, was just such a law gave rise to an often bitter debate on population growth theory which was only subsiding by the time of Pearl's death in 1940.

Pearl and Reed's equation described the growth of a population along a smooth, S-shaped path toward a stable upper limit (Figure 3.1). They were unaware at the time that a Belgian mathematician, Pierre-François Verhulst, had actually used the same equation to describe population growth over eighty years earlier. Verhulst had become interested in the problem through his mentor, Adolphe Quetelet, who in 1835 had proposed that the resistance to the growth of a population was proportional to the square of the speed with which the population tended to increase.[44] This principle appealed to Quetelet because he saw in it a direct physical analogy with the resistance of a medium to a body traveling through it. He suggested that Verhulst submit the principle to examination and compare it to available population data.

Verhulst's subsequent analysis forced him to admit his failure to find a law of population, because he was unable to determine the exact nature of the function that described the obstacles to growth.[45] Nevertheless, he was able to obtain a theoretical, that is, mathematical, solution to the problem by operating on the assumption that the rate of growth was retarded by a function linearly proportional to the size of the "superabundant" population. The superabundant population referred to the number in excess of the population existing at the moment when resources, in this case farmland, became limiting, this latter number being called the "normal" population. His analysis yielded an S-shaped curve of growth over time, which he labeled the "logistique."[46] In fact, the mathematics of his derivation did not correspond to his verbal reasoning, for there was no justifica-

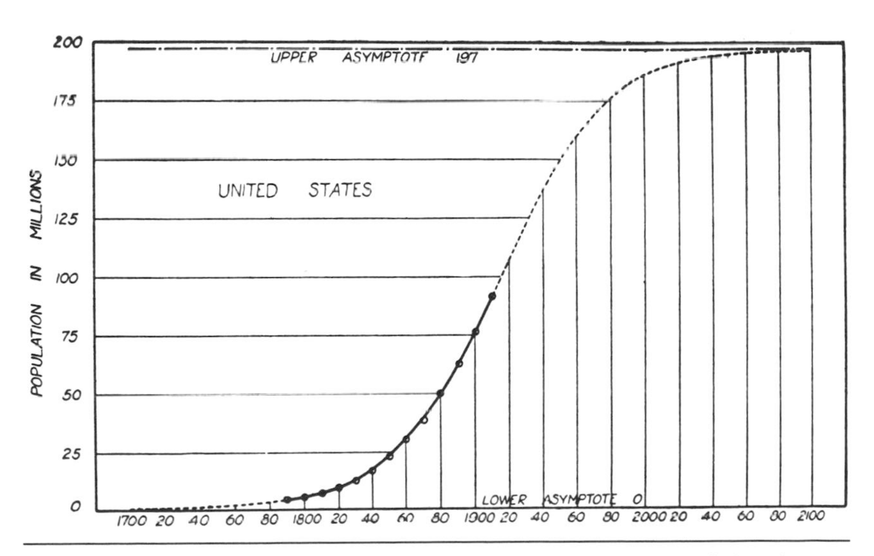

FIGURE 3.1. The logistic curve fitted to U.S. census data, but not including the 1920 census results. Over half of the curve is extrapolated. (From R. Pearl, *The Biology of Population Growth*, New York: Alfred A. Knopf, 1925, p. 14.)

tion for arguing that a population shifted from normal (exponential) growth to superabundant (logistic) growth at some crucial density: his symmetrical logistic curve did not reflect this sharp transition.

Verhulst did not explain his choice of the term "logistique" for his curve, but in nineteenth-century French the word referred to the art of calculation, as opposed to "theoretical arithmetic";[47] it was applied to a type of logarithm used for astronomical calculations. From the context of Verhulst's 1845 memoir, it is likely that he intended to convey the idea of a calculating device, from which one could calculate the saturation level of a population and the time when it would reach that level. In any case, he realized that the logistic curve was only one possibility; the obstacles to growth could also be proportional to the square of the superabundant population, for instance.[48] In his last memoir of 1847, he suggested that the obstacles were proportional to the ratio of the superabundant to the *total* population.[49] Verhulst died in 1849, two years after his memoir, and his population work remained largely unnoticed. Quetelet, who might have publicized the results, strongly disagreed with Verhulst's bold conclusions, which not only failed to support Quetelet's principle, but lacked any physical analogy of the sort that he favored.[50]

Eventually Pearl and Reed discovered Verhulst's 1845 memoir and adopted the term "logistic" for their curve.[51] However, the true precursor of their work was not Verhulst, but T. Brailsford Robertson, a physiologist who in 1908 published two articles in which he applied the same sigmoidal curve to various cases of individual growth in animals, plants, and man.[52] By coincidence, he used some of Quetelet's data without being aware apparently of Quetelet's population interests or of Verhulst's memoirs.[53] Robertson called his curve the "autocatalytic" or self-accelerating curve, because it was identical to that used to describe a certain type of chemical reaction in which one of the products of the change had the property of accelerating the further progress of the reaction. Strictly speaking, the term "autocatalytic" should refer only to the accelerating part of the curve, but Robertson used it to describe the whole S-curve. This loose designation was repeated by others, with the result that the logistic and the autocatalytic curve came to be synonymous. On the basis of this similarity between chemical and growth phenomena, Robertson constructed an elaborate hypothesis to explain the changes observed in the growth rate of individuals. He postulated that growth itself was an autocatalytic phenomenon, controlled by the secretion of an unknown catalyst which would act to stimulate growth. He later suggested that the autocatalyst might be the phospholipid lecithin.[54] The original theory applied only to the growth rate of individual organisms, but in 1923 Robertson extended his hypothesis to protozoa and bacteria populations.[55]

In 1909 Raymond Pearl wrote a review of recent growth studies in which he sharply criticized Robertson's use of the autocatalytic curve. Finding fault with Robertson's curve-fitting procedure, Pearl commented that the discrepancies between observation and theory were "so great in amount, so biased in character and so frequent in the data presented that these data, as they stand, cannot reasonably be held to afford evidence of any particular value in favor of Robertson's ingenious, suggestive and potentially very valuable hypothesis."[56] Robertson's curve was not the one that Pearl himself had used to describe growth in an earlier study, started while he was teaching at the University of Michigan in 1903, on the growth of the aquatic plant *Ceratophyllum demersum*. Pearl published this work in 1907 after his year in London with Karl Pearson, whose biometric methods he applied in this early study.[57] His lack of enthusiasm for Robertson's curve probably reflected his faith in his own curve, a logarithmic equation relating the number of leaves on a whorl (y) to the position of the whorl on a primary branch (x):

$$y = A + c \log (x - \alpha). \tag{3.1}$$

In this equation, Pearl thought that he had found the mathematical law of growth in *Ceratophyllum*.

A second criticism was based on Robertson's use of the curve to support his hypothesis of growth. Pearl pointed out that similarity between curves implied nothing about the underlying mechanisms of growth. What was required was to show that there were qualitative and not merely quantitative resemblances between the two kinds of phenomena. He argued that "similarity of quantitative relations between phenomena cannot safely be taken as proof . . . of qualitative identity, because of the observed general lack of uniqueness in the quantitative relations of natural phenomena."[58] These remarks are of interest not only because they illustrate Pearl's cautious approach to growth curves in 1909, but also because, some years later, both of these criticisms were to emerge from the pens of Pearl's equally righteous critics. The fact that they could reappear indicates how strongly Pearl had become attached to his logistic hypothesis.

In their 1920 article, Pearl and Reed applied Robertson's autocatalytic curve to census data for the United States on the basis of certain assumptions which they thought must hold for any population (see Figure 3.1). Given a limited area into which a population could expand, they argued that the rate of population increase at any time was proportional to two things: the magnitude of the population at that time, and the "still unutilized potentialities of population support existing in the limited area."[59] The meaning of "unutilized potentialities" was vague. Pearl later expressed it as the "amount still unused or unexpended in the given universe (or area) of actual or potential resources for the support of growth."[60] This version was only slightly more satisfactory, for it left the problem of trying to estimate "potential resources." A sympathetic commentator offered a simpler interpretation of "unutilized potentialities" as "the difference between the existing and the limiting population."[61] Despite the fact that Pearl and Reed did not express it this way, this version reflected the meaning of their assumption more accurately. The rate of growth, therefore, was proportional to two quantities: the existing population and the difference between existing and limiting populations. What Pearl and Reed had essentially done was to begin with the assumption that the population would follow an S-shaped curve, and they later admitted as much.[62] Therefore they had assumed that which had to be proved, yet they presented their curve as being empirically fitted, not logically derived. Finding the right mathematical formula was a simple problem given the familiarity of Robertson's autocatalytic curve.

Pearl's conclusion that his curve was a law of growth was merely an extension of Karl Pearson's view of science as a process of formulating scientific laws through the continual refinement of hypotheses. A law of this kind did not have to provide any explanation in the sense of referring to any underlying mechanisms; it just had to describe growth reasonably accurately in past and future. Pearl felt that his curve did describe past events, and that because of the rational assumptions on which it was based, it would give a good picture of future events. Other curves such as the Malthusian or exponential curve could not be called laws because they did not depict the whole history of growth, leading to absurd values at some point.

However one may want to define the meaning of scientific law for the purpose of philosophical discourse, in practical usage the term "law" has been loosely applied to many different kinds of statements. As A. D. Darbishire noted in 1906, the word "law" was in biology the vaguest of all terms, "signifying as occasion demands either a theory, or a résumé, or a hypothesis, or a formula, or a generalization," and sheltering "under its wide roof, Laws whose authors aim at explanation, and those whose authors are satisfied with description."[63] Therefore to announce that one has uncovered a law may not mean very much at all. But it is also the case that the use of the word "law" sets off emotional resonances. In announcing that we have found a law, we also suggest that we have discovered a truth of some importance; that if we have not gone so far as to ape Newton, we have at least aped Boyle. Pearl did feel this sense of achievement: he compared his law "in a modest way" to Kepler's laws of planetary motion and to Boyle's law.[64] As he bent through his successive mental contortions trying to uphold the statement that the logistic curve was a law of growth, he managed to convert a fairly minor issue into a major controversy. Let us now look at what Pearl did with his curve.

The Logistic Curve in Theory and Fact

The validity of Pearl's curve as a law of growth rested on its ability to describe the available data. But it was immediately apparent that not all populations showed smooth sigmoidal growth. In this respect, it was clear that to achieve the required correspondence between observation and theory would entail some modification of the theory. Accordingly Pearl and Reed regarded their equation as a first approximation of the law, not to be taken as having exact predictive value. Their caution, however, did not prevent them from giving a projected figure for the limiting population of the United States of around 197 million (to be attained just after the year 2000), nor from estimating the point of inflection of the curve down to the exact day: 1 April 1914.[65]

The most obvious problem was that a single logistic curve represented only one cycle of growth, covering a definite time span. The actual length of a cycle was not fixed but depended on the available data; for the United States it covered roughly 200 years. In order to account for longer periods, Pearl and Reed proposed that growth occurred in successive cycles, each one represented by a logistic curve.[66] A new cycle would begin when a major change, such as an industrial revolution, created the opportunity for growth beyond the limiting value dictated under the existing system. T. B. Robertson had also used successive cycles of growth in his own work and had been criticized by Pearl on the grounds that he had not proved conclusively that growth was in all cases cyclical in character.[67] Pearl and Reed completely ignored this objection as applied to their own case and simply assumed, although there was no reason to do so, that population growth would be cyclical—one logistic curve added onto the next through time.

The second problem was more serious, for it involved the symmetry of the curve as they had originally presented it. Symmetry meant that the inflection point came at the halfway point of the curve, and that the saturation population was exactly twice the population at the point of inflection. Pearl and Reed tried to justify their doubts that the curve was in reality symmetrical by arguing that symmetry implied that the forces acting to inhibit growth in the latter half of the curve were equal in magnitude and exactly similarly distributed in time to the forces which operated to accelerate growth in the first half. "We do not believe that such rigid and inelastic postulates as these are, in fact, realized in population growth," they added.[68] Since the meaning of this implication of symmetry was unclear as long as these forces were undefined, it is hard to accept that they were really so troubled by this problem. A different but unstated possibility is also likely: Pearl and Reed's reluctance to limit themselves to a symmetrical curve may have stemmed from their desire to make their law universally valid, no matter what kind of growth was being considered. A symmetrical curve was clearly not adequate to cover all the data, as Pearl had pointed out with reference to T. B. Robertson's work. Indeed, their first attempt to use an asymmetrical form of the curve was to refit some of Robertson's own data on individual growth, which he had earlier tried to fit to a symmetrical curve, with poor results (Figure 3.2).[69]

In the process of freeing the logistic curve from its restrictive symmetry, Pearl and Reed generalized their original equation by adding more terms to it. Thus the original equation could be written in the following form:[70]

$$N = \frac{K}{1 + me^{at}}, \qquad (3.2)$$

where N is the number of individuals; t is the time; e is the base of natural logarithms; and K, m, and a are constants. The generalized equation was then:

$$N = \frac{K}{1 + me^{a_1t + a_2t^2 + a_3t^3 + \ldots + a_nt^n}}. \qquad (3.3)$$

The actual shape of the curve would depend on the number of terms and the value of the constants. If all constants from a_2 up to a_n were zero, the

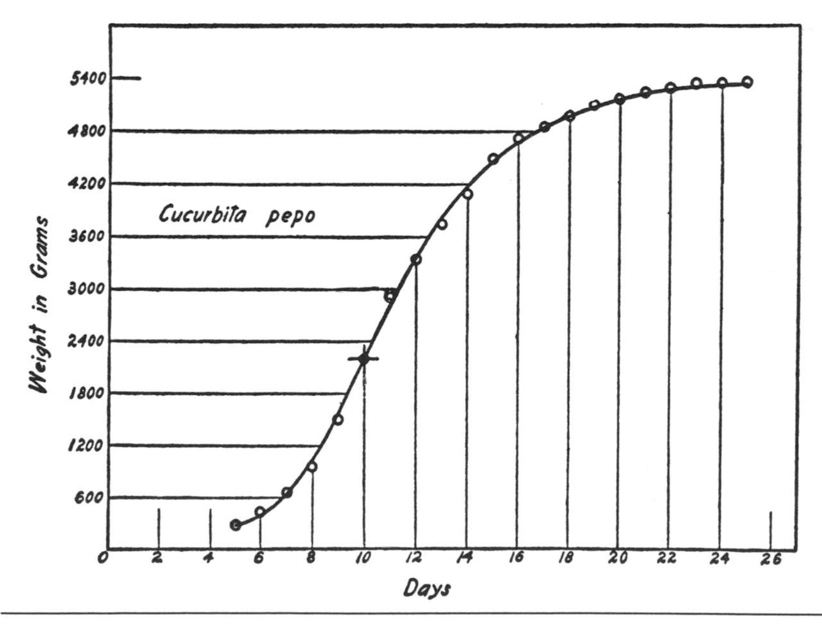

FIGURE 3.2. Asymmetrical logistic curve, used to represent the growth of the pumpkin, with data taken from T. B. Robertson. (From R. Pearl, *The Biology of Population Growth*, New York: Alfred A. Knopf, 1925, p. 7.)

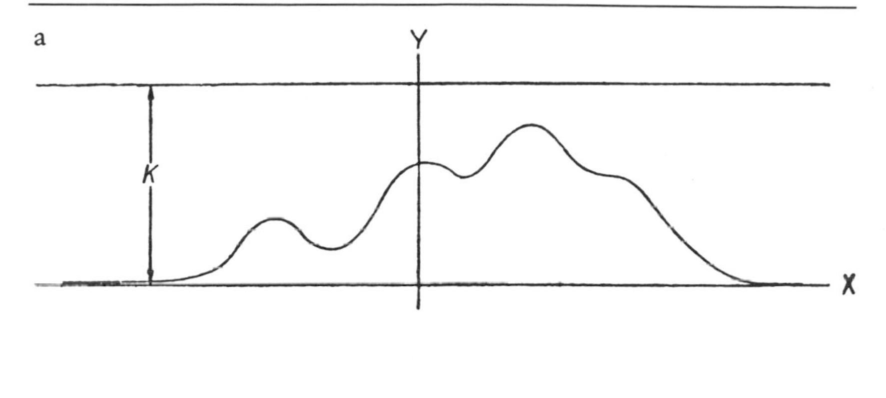

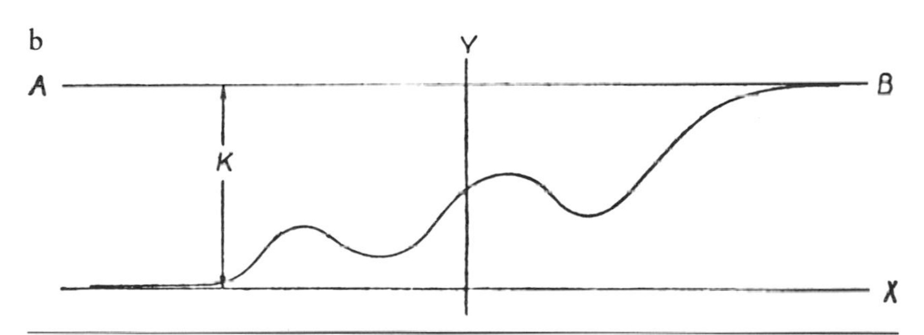

FIGURE 3.3. Pearl's illustration accompanying his discussion of his generalized equation. (*a*) form of the curve when n is even and a_n is positive; (*b*) form when n is odd and a_n is negative. (From R. Pearl and L. J. Reed, *Metron* 3 [1923], 15.)

curve would reduce to the original symmetrical form. If the terms up to the x^3 term only were left in, the curve would have an asymmetrical sigmoid shape. Otherwise, a variety of shapes was possible (Figure 3.3). Apart from the difficulty of determining so many constants, the problem with this type of generalization was that it could be made to fit almost any set of data and was therefore hardly the calculating device implied by the term "logistic." Pearl was fully aware of this type of error, for he warned against it in 1923 in a textbook on statistics written for medical students:

> [The experienced person] knows that by putting as many constants into his equation as there are observations in the data he can make his curve hit all the observed points exactly, but in so doing will have defeated the very purpose with which he started, which was to emphasize the law (if any) and minimize the fluctuations; whereas actually if he does what has been described he emphasizes the fluctuations and loses completely any chance of discovering a law.[71]

But his enthusiasm for the logistic curve was such that he failed to perceive the aptness of this remark to his own methods, and he believed that in the expanded equation he had succeeded in setting forth a "comprehensive general theory of population growth."[72] In fact, however, Pearl and Reed seldom had recourse to the generalized curve beyond the simple asymmetrical S-shaped form, except as a final, if inadequate, rebuttal to criticisms that populations did not always follow a "logistic" curve.[73]

Having presented the general theory, it remained to test it against the data. In 1924 Pearl and Reed fitted census figures of sixteen countries, the

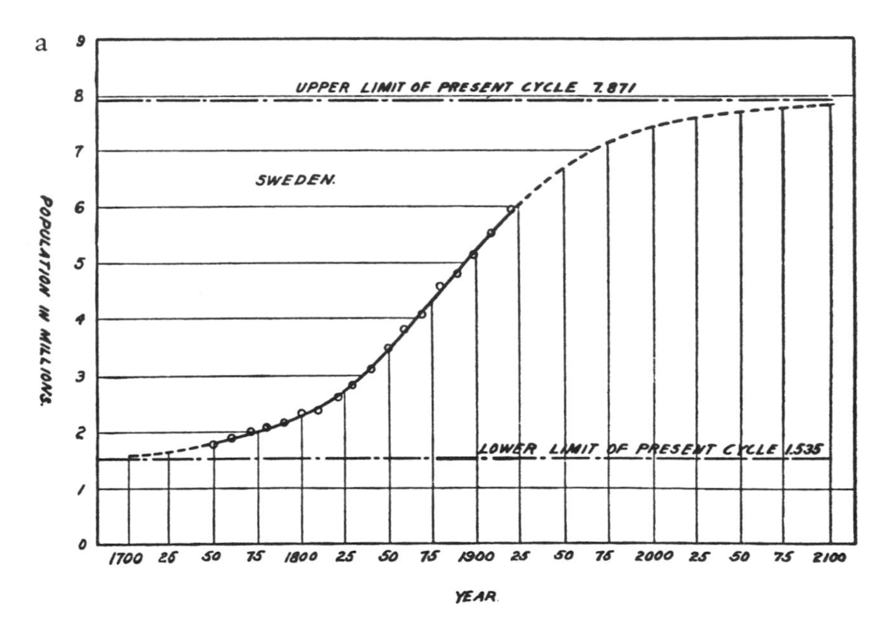

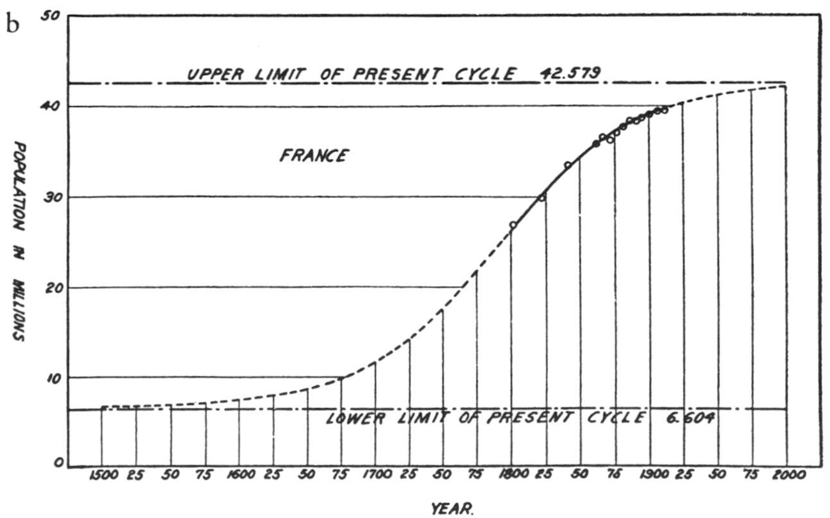

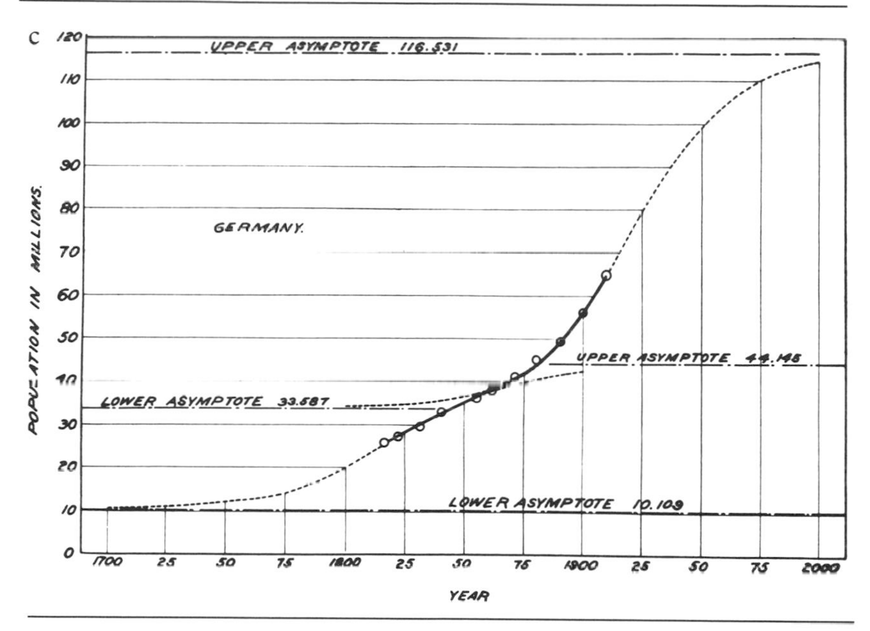

FIGURE 3.4. The logistic curve fitted to three populations: (*a*) Sweden; (*b*) France; (*c*) Germany. Large sections of the curves are extrapolated. Germany is here fitted to two curves, one growing out of the other. (From R. Pearl, *The Biology of Population Growth*, New York: Alfred A. Knopf, 1925, pp. 12, 16, 21.)

world as a whole, and one city to logistic curves (Figure 3.4).[74] In only one case, however, that of the city of Baltimore, were the data sufficient to justify the application of the logistic curve, a consideration which Pearl and Reed simply disregarded. In their first paper of 1920, the U.S. census data covered less than half the curve, not even reaching the inflection point, since they had not used the 1920 census results. The entire upper half of the logistic was therefore extrapolated.[75] For Sweden, where population records went back the furthest, the observations covered just over half a cycle, despite Pearl and Reed's claim that they covered "roughly two-thirds of the entire cycle."[76] For Germany and Japan, two cycles had to be invented in order to fit the data, on the justification that both of these countries had undergone major industrial changes in the recent past. The observations available on Denmark filled less than a third of the curve, yet they remarked that "the fit of theory to observations is well-nigh perfect."[77] To be sure, the fits were nearly perfect in the sense that all the points lay close to the curve, but this did not diminish the fact that their attempts to fit logistic curves to all these countries were largely arbitrary.

Pearl and Reed did not rely solely on census data to prove the correctness of their curve, however. They assembled figures from other sources showing that growth in general, whether on the individual or population level, followed a logistic path. As a result of T. B. Robertson's work, the autocatalytic curve had been applied to a wide variety of cases of individual growth by the mid-twenties.[78] The same curve had been used independently by other biologists to describe the growth of microorganisms and yeast populations as early as 1911. A. G. McKendrick and M. Kesava Pai had arrived at the equation by assuming that the rate of increase was proportional to the number of organisms present and to the concentration of food.[79] Pearl and Reed both cited and reproduced the results of these studies as support of their logistic hypothesis. In his own laboratory, Pearl and his assistants made use of the logistic curve whenever possible to describe the growth of the fruit-fly populations they were studying.[80]

It is significant that Pearl and Reed differed from previous workers in the way they wrote the equation, however. It had been the practice to write the logistic equation in two forms. One related the number of individuals to time:

$$N = \frac{K}{1 + e^{a-rt}}, \tag{3.4}$$

where N is the number of individuals; t is the time; r is a constant giving the maximum, unrestricted growth rate; K is a constant representing the upper limiting population; and a is a constant of integration. The graph of this curve is the symmetrical, S-shaped curve of population growth.

The second way to write the equation is in its differential form, where the rate of growth is expressed as a function of the number of individuals at time t:

$$\frac{dN}{dt} = rN\left(\frac{K-N}{K}\right), \tag{3.5}$$

where dN/dt is the rate of increase of population, and the constants r and K are the same as in the integrated equation (3.4). What is worthy of note is that, while the graph of the first equation shows most clearly the history of a population over time, it is actually the second statement of the logistic curve—the differential form—which is easier to interpret and to analyze. As I shall show in chapter four, it is fairly easy to derive the differential form after making several simplifying assumptions about how populations must behave.

Most of the people who used the logistic equation before Pearl, did express it in its differential form, at least to begin with. Verhulst in 1845 used a different line of reasoning, but his reasoning led him to the differential equation, which he then integrated to get his logistic curve. T. B. Robertson based his curve on a chemical analogy which was open to much criticism, but he also began by writing the differential expression. McKendrick and Pai reasoned toward the differential equation first and then expressed the integrated form of the curve. Indeed, Pearl stands out as an exception in his consistent use of the integrated form of the equation. He did use the differential equation in the course of some arguments, for example, when he showed how the generalized equation could be derived from the simple curve, but he never actually derived the differential form directly from his assumptions, as did nearly everyone else.

In view of the fact that Pearl was familiar with these other studies, it seems strange that he should have avoided the differential expression, particularly since he did adhere to the basic assumptions underlying the curve. Part of the reason was his interest in forecasting human population trends; for this purpose the S-curve is most useful, and it is reasonable that he would have written the equation corresponding to that curve. But it is also possible that his reluctance to use the differential form was connected to the obviously arbitrary nature of its derivation. As Verhulst had concluded, the logistic equation, while mathematically interesting, was not the law of growth. Pearl's reliance on the more complicated integrated form, even in the *Drosophila* work, could also be seen as a way of emphasizing the ability of the curve to fit the data, while deflecting any criticisms aimed at the underlying assumptions.

By 1924 Pearl had accumulated enough examples of the use of the logistic curve that he could confidently proclaim, "In the matter of population growth there not only 'ought to be a law' but five years' research has plainly shown that there is one."[81] By 1927 he asserted, "It is an observed *fact*, which at this stage of the discussion involves no theoretical implications whatever, or postulates special to it, that the growth of populations of the most diverse organisms follows a regular and characteristic course."[82] Pearl had come full circle: having first *assumed* logistic growth in trying to find a curve to fit his initial data, he now believed that the empirical evidence proved the truth of the logistic "law," even though large parts of the curves were extrapolated.

Pearl understood that new ideas gained strength through repetition, and he did what any shrewd man would do to ensure that an idea not pass unnoticed: he launched a crusade. From 1920 to 1927 he published, either alone or with Lowell Reed, over a dozen articles devoted to the logistic curve, which appeared in an array of journals designed to make his discovery accessible to the largest audience.[83] These included his own journal, *The Quarterly Review of Biology*, and the journal of his good friend H. L. Mencken, *The American Mercury*. Although a few of the articles dealt with specific modifications of the theory, for the most part they were almost identical statements of the basic theory and application of the curve.

This flurry of publications was supported by several seminar presentations. In 1925 he and Reed took part in a meeting in New York devoted to the problem of forecasting populations, where discussion centered on the logistic curve.[84] Pearl estimated the audience at 250–300 people, although the published report of the meeting listed only 147 people present. "On the whole," wrote Pearl to a friend, "I should say that honors were in our direction overwhelmingly. There were but few dissenters and they were not people of heavy weight."[85] George Udny Yule, British statistician and Pearl's friend from his early days with Pearson, presented Pearl's theory in a talk delivered at the meeting of the British Association for the Advancement of Science held in Toronto in 1924.[86] The talk, expanded and published the following year as Yule's presidential address to the Royal Statistical Society, provided a complete discussion of the theory as developed both by Verhulst and by Pearl and Reed.[87] Yule emphasized that the curve could not be used for long-term prediction due to the uncertainties caused by wars and other disruptions, but he was otherwise very supportive of Pearl's work. His article, which Pearl cited often in later papers, was important in explaining and lending stature to the logistic theory.

Pearl himself discussed his theory in a session of the World Population Conference held in Geneva in 1927, later publishing his paper in his journal.[88] Finally, the logistic curve was a prominent feature of three books that Pearl published during this period: *The Biology of Death* (1922), *Studies in Human Biology* (1924), and *The Biology of Population Growth* (1925).[89] With this prodigious flow of words, Pearl's logistic curve attracted a great deal of attention and, inevitably, a great deal of criticism.

Much Ado

First Skirmishes

The first to scrutinize the logistic curve were economists and statisticians. The kinds of criticisms voiced by statisticians in particular reflected the strong empirical bent of the discipline, which fostered a conservative attitude toward generalizations of any kind. Demographers were in the business of making predictions, which then became the basis of specific planning proposals; because demography was an applied science, the statistician tended to view his ability to predict with great caution. Although it was necessary to extrapolate from the data in order to estimate conditions between census years, there was little motivation to attempt any extrapolation beyond the next census. The very nature of this work, its attention to detail, instilled certain attitudes born of an acute awareness of the complexity of events and the concomitant hesitation to venture far beyond the facts. For this reason the most frequent criticism of the logistic curve was that there were too many uncertainties even to attempt to use a regular curve like the logistic, let alone to have any confidence in its predictions.[1] The logistic curve was not a useful curve.

Pearl did feel his curve was a reliable predictor of long-term trends, but he was loath to have it judged on that basis alone. In an effort to forestall the above objection, he went into a remarkable about-face.

> The extrapolated portions of the curve have no bearing whatsoever upon the adequacy of the curve to describe the known *facts* of population growth. It is upon the success or failure of the attempt to demonstrate this adequacy, *and upon this alone*, that the judgment of the scientific validity of the hypothesis must rest.[2]

These remarks merely fueled his critics' fires. Expressing the economist's point of view, A. B. Wolfe pointed out that such a statement was hardly in accord with Pearl's insistence that his curve was a law of population growth, implying that it did reflect the underlying mechanisms of growth: "If it somehow reflects the mechanism—a mechanism explainable or describable in terms of known principles—by which human population growth is what it is, it should also reflect what that growth will be."[3] And if the curve did not after all have any long-term predictive reliability, then any other empirical cirve already in use for short-term prediction was equally useful.

Again to forestall the same objection, Pearl stressed that the logistic curve could be used only on the assumption that no fundamentally new factor influencing the population should come into play in the period under observation.[4] Arguing that it was impossible to make this kind of assertion with respect to human populations, where social, economic, political, or religious changes were highly unpredictable, Wolfe accused Pearl of not taking his own proviso seriously enough:

> One could be tempted to the inference that Pearl is driven by the fine frenzy of pure intellectual play, did he not in his later writings broadly hint that he feels himself to be on the trail of a great discovery—a "rational," "mathematico-biological" law of population growth, universally valid.[5]

A similar objection was raised by George H. Knibbs, a demographer whose monumental analysis of the 1911 Australian census had failed to disclose any general law of population growth.[6] The key to predictability for Knibbs lay not in speculative graph work, but in the analysis of the mutual reactions of social and economic factors on the rate of population increase. He suggested that more exhaustive studies, of the type he had made on the Australian census, were what was needed.

It was not long before Knibbs felt compelled to add his voice to the rising chorus of criticism against Pearl's methods.[7] His attacks were based on what he perceived to be an erroneous assumption of the theory: that the "reproductive impulse" of a population was constant through time, but was prevented from expressing itself completely by the effects of population density and resource limitation. He argued that other kinds of factors, such as social, ethical, or economic changes, could lead to an intensification or diminution of the "impulse to increase." He meant by this that the group's standard of living, including its entire attitude toward reproduction, was determined by its social and ethical outlook. Therefore a change in the basic "character" of the people, acting independently of the population density, could cause changes in the rate of increase. Whatever the results of this change, it could not be subject to law. The use of the logistic curve was perceived as a misleading exercise in curve fitting which could not be considered as proof of the theory.

This issue was further complicated by the feeling that adherence to the logistic curve implied a degree of complacency with respect to population growth. Knibbs, for instance, believed strongly that Malthusian growth would lead to dire consequences unless checked by immediate social measures, which in this case meant a eugenics program designed to raise the level of human intelligence, for the ultimate purpose of controlling the rate of increase.[8] The logistic theory seemed to lead to the conclusion that one need no longer feel a sense of urgency in these matters. Indeed, Pearl did at this time begin to relax his views on eugenics. Previously he had been a strong advocate of eugenic principles, following the lead of Karl Pearson. In the late twenties he softened his position and began to argue against the class bias that pervaded the eugenics movement.[9] He still believed that heredity rather than environment played the main role in producing intellectual superiority, but he suggested that it was not undesirable to have the lower classes reproducing faster than the upper, since superior people were more often found to arise from groups humbler than the aristocratic. These views prompted further attacks from eugenicists. Pearl made no secret of the fact that he felt the eugenics movement contained too much sloppy thinking, which indeed it did, and he argued that it was time to replace eugenics with a more scientific approach aimed at human genetics. Partly with that goal in mind, he helped to create the International Union for the Scientific Investigation of Population Problems (later the International Union for the Scientific Study of Population) after the first World Population Conference held in Geneva in 1927. The union was meant to act as a coordinating body for population research conducted in a "strictly scientific spirit."[10]

Similarly, Pearl's theory, by suggesting that decrease in growth was an automatic response of a population growing to a certain level, implied that birth control had no effect on the modern population decline. The "contraceptive hypothesis" was therefore advanced as an alternative to Pearl's hypothesis.[11] These complaints were reasonable on the part of Pearl's critics, for he did conceive population growth to be a biologically self-regulated process, without specifying what he meant by self-regulation, and he did ultimately cease to regard population growth as a cause for alarm. However, he was also an advocate of birth control, which he regarded as an "intelligent adaptive response to an environmental force, population pressure,"[12] and therefore one of the reasons the population did follow the logistic curve. But the critics quite rightly did not accept the inherent circularity of Pearl's thinking.

For those who had been accustomed to view the complexity of human society as proof that growth could not be reduced to a simple law, Pearl's apparently casual neglect of environmental and social factors seemed wholly wrongheaded. Pearl did believe that major changes such as industrial revolutions had a measurable impact on growth, and he did not hesitate to appeal to such major changes when he needed to postulate two or more cycles of growth to fit a given set of data. But in most cases he believed that nothing short of catastrophic turmoil would shift a population from its logistic path; even wars and epidemics made "the most ephemeral flicker in the steady onward march of population growth."[13] Therefore all that was necessary was to set down the law of growth. For the critics, less imbued with Pearsonian positivism, this could not serve as a satisfactory explanation of anything.

Eventually Pearl's manner of expression, if not his attitude, altered in response to these criticisms. In his earlier writings he tended simply to exclude the actions of external circumstances from consideration with respect to the logistic curve, but by 1927 he was arguing that the equation did indeed take these factors into account. Their effect was not measured independently, however, but in terms of what he called the "primary biological factors" of natality, mortality, and migration (migration in this case being considered unimportant). The logistic curve included environmental variables "in the sense that it describes the integrated end effect upon population size of the aggregated forces tending towards increase on the one hand, and decrease in numbers on the other hand."[14] Ignoring the fact that his use of the curve reflected assumptions not yet proven, he answered the arguments of critics who claimed that he had neglected environmental factors with a vigorous dismissal:

> This argument is rubbish, born out of the conservative resistance to any new idea which the established order of learning has always shown, by that wind-broken and spavined old stallion, faith in *a priori* logic as against plain facts of experience.[15]

The vehement tone adopted by Pearl and his critics during this debate reflected not only a difference in attitude over what may be called a law, but also a difference in the way the population itself was seen. Underlying Pearl's argument was the assumption that the population was a discrete entity, loosely analogous to the individual organism. This idea of population was not shared by Pearl's critics, nor was it made explicit by Pearl himself in his writings before 1927. Much of the gap in communication between Pearl and his critics arose from Pearl's failure to make sufficiently clear just how he regarded populations and on what assumptions this idea was based.

The idea of the population as a discrete entity tied in with Pearl's work in 1907 on the growth of aquatic plants and with his lifelong interest in hereditary versus environmental factors influencing the individual.[16] In the 1907 study he had used a logarithmic equation as a "law" of growth in the plant. But in addition to providing a simple description of growth, he had used the equation to draw a basic distinction between the internal (genetic) factors influencing growth and the external (environmental) factors. In other words, Pearl perceived two kinds of factors operating independently on the growth of the plant which he called "internal form-determining factors" and "external environmental factors." These two categories of factors were represented by different parts of the growth equation; moreover, it was the "internal" portion of the equation that gave the curve its *shape*; the "external" portion determined only its position on the *y*-axis. For population growth the equation differed but the principle was the same: Pearl assumed a similar, but in this case more artificial, distinction between the "biological" (internal) and the "physical" (external) environment of the population. This separation between kinds of factors arose because he perceived the population essentially as he had perceived the individual organism, and because he still believed that a clear distinction could be made between the effects of hereditary (internal) and environmental (external) factors. His whole approach to the analysis of growth remained virtually unchanged from what it had been twenty years before in his work on individual growth.

"Populations of whatever organisms are, in their very nature, aggregate wholes, and behave in growth and other ways as such," he wrote in 1927.[17] The assumption behind this statement, that the population behaved like the individual organism, meant that it was permissible to examine growth exclusively in terms of the "biological" environment, that is, in terms of the attributes, such as birthrate and deathrate, that belonged to the population itself rather than to the external world. The "physical" environment, with its intractable variables, could be conveniently given second place.

The fact that both individual organisms and populations followed the same logistic path was circumstantial evidence that they were fundamentally alike. But, as Pearl had so vehemently argued in response to T. B. Robertson's work, and as he continued to stress, quantitative evidence was not sufficient to prove qualitative similarity. In order to prove that a population did behave like an individual, he needed to show that the shape of the growth curve could be explained by some mechanism "internal" to the population, that is, a mechanism involving the intrinsic "biological" attributes of the aggregate. He thought he had found this evidence in his studies of the fruit fly, *Drosophila*.

Starting in 1921, Pearl and the associates in his laboratory had been conducting an intensive series of investigations on the duration of life, using different strains of the fruit fly as experimental subjects. In the course of their experiments, which were conducted on small laboratory populations of flies enclosed in milk-bottle-sized universes, it soon became apparent that the density of the population might influence the duration of life of the flies. The finding of a definite relationship between deathrate and density came as no real surprise; previous statistical work indicated a similar relationship in human populations.[18]

More controversial was Pearl's claim that density was also affecting the birthrate of fly populations by inhibiting fecundity, or the number of eggs laid per female.[19] Pearl found that there was a regular decrease in the rate of reproduction with increasing density of population. The smoothness of the curve describing the relationship indicated to Pearl that this was not a haphazard effect, but a "highly lawful phenomenon," a causal connection between reproduction and density. Here, Pearl felt, was a solution to the problem of the mechanism behind the logistic curve, a *vera causa* to explain why the rate of growth decreased in the upper part of the curve.[20]

He also noticed a similar connection between density and egg laying in populations of Plymouth Rock fowl with which he had worked more than a decade earlier. Impressed by these results, he then went a step further, claiming in 1925 to have demonstrated that human populations also showed a definite density effect on birthrate. With scant data, he obtained by partial correlation analysis a small negative correlation ($-.175 \pm .052$) between birthrate and density in a human population, and he suggested that the relationship was of the same fundamental character as that found in hens and flies.[21]

His critics viewed this connection between density and rate of growth as somewhat mystical, in particular when applied to humans. As A. B. Wolfe remarked, "So small a coefficient is entirely unequal to the burden put upon it."[22] British social biologist Lancelot Hogben discounted the idea that fruit fly and human populations could be compared in this manner, and in so doing echoed Pearl's own criticisms of T. B. Robertson in 1909:

> The mere fact that the same type of equation can be used for two different sets of variables does not necessarily denote the same intrinsic mechanism. . . . Pearl's conception of density, as applied to human populations, remains a purely statistical abstraction devoid of specifiable biological significance. The mere fact that an assumption can be tested in a biological laboratory does not of itself confer on it the status of a biological hypothesis. If, and only if, its biological significance can be clearly envisaged, is it permissible to assign any theoretical meaning to a correspondence between the growth of human populations in successive cycles and the single cycle experimental curves of *Drosophila* or yeast.[23]

The problem of establishing the biological significance of these results was an important one. Pearl had first published these conclusions, with his heroic extrapolations to human society, without any suggestion of how density might influence reproduction. Rather belatedly, in 1932, he tried to remedy this flaw in a study of the mechanics behind the density effect he thought he had uncovered in *Drosophila*.[24] Using the analogy of gas molecules colliding randomly in an enclosed space, he interpreted the reduced fecundity as an interference effect of the flies on each other which altered three activities: food intake, energy output in muscular activity, and oviposition. This explanation, relating the "biological" attribute of the aggregate (density) to physiological processes of the individuals, fitted in well with Pearl's concept of the population as a discrete entity and with his belief that there were grounds for comparing human and animal populations. It was also compatible with his idea that the logistic equation was a law comparable to Boyle's law: the density effect, based on the movement of the individuals, was comparable to the way the kinetic theory of gases supported Boyle's law. These experiments were valuable for showing that density could act through its effect on behavior, but Pearl's error was to suggest that the effect clinched his case that the logistic equation was the law of growth.

Despite Pearl's claims, he had not demonstrated why his curve deserved the status of a law of growth. Economists, demographers, and biologists alike were justified in remaining skeptical in the light of Pearl's haste to employ analogies, based on meager evidence, between individual organisms and populations, and between human and animal populations. One by one they pierced the many vulnerable places of Pearl's armor, calling attention to the factors he had not included, the interpretations he had not considered, and the analyses he had not performed. Not all the criticisms were delivered in so vehement or negative a way as those discussed above, however. Within Pearl's own orbit, Alfred Lotka, in the course of writing his book, had turned his attention also to the logistic curve and was developing a more evenhanded assessment of its potential. His approach involved seeing the logistic curve not as a universal law in Pearl's sense, but as an "animated question mark."

An Animated Question Mark

Lotka had been interested in population growth for several years without, however, using a sigmoid growth curve, despite the fact that he knew of T. B. Robertson's work at least as early as 1910.[25] One of Lotka's earliest studies in 1907 dealt separately with examples in which growth was exponential and examples in which the population was stationary, but he did not combine the two into a single equation including both a rapid increase and an equilibrium phase.[26] In later papers, prior to his book of 1925, Lotka assumed that growth was basically exponential.

Contact with Pearl failed at first to excite his interest in the logistic curve, although Lowell Reed recalled that he and Lotka would discuss these problems often at lunch, leaving the restaurant tablecloth covered with drawings of exponential and logistic curves.[27] As Reed explained, "Lotka at that stage had no use for the logistic for to him it was merely another empirical equation."[28] His opinion changed dramatically in the summer of 1923, when he discovered that the logistic curve was the solution to one of his problems in population theory. Reed described the conversion:

> The next summer when the members of the Department were at work in a small summer laboratory in central New Hampshire, Dr. Pearl received an

> enthusiastic letter from Lotka; he had just noted that the logistic was the solution for one of the simple cases of his family of differential equations and his attitude with regard to the curve had changed completely. From that time on . . . he developed a great deal of his theory of the dynamics of population using the logistic as the basic function.[29]

Reed's statement errs in giving the logistic curve a much larger role in Lotka's work than it actually had; much of his analysis began with the postulate of exponential, rather than logistic growth, which he then modified in various ways. Lotka's contribution to logistic theory was restricted to his demographic papers in which he explored in detail what it meant for a population, in terms of its demographic traits, to be growing logistically.

In *Elements of Physical Biology*, Lotka considered, as a special case of the kinetics of evolving systems, a single population growing under constant conditions.[30] The growth of the population was assumed to be some function of the population, X:

$$\frac{dX}{dt} = F(X). \tag{4.1}$$

The problem was to determine the values at which equilibrium would be established, that is, when $F(X) = 0$. Using Taylor's theorem, he expanded the equation, omitting the initial absolute term so that dX/dt would vanish with X:

$$\frac{dX}{dt} = F(X) = aX + bX^2 + cX^3 + \dots. \tag{4.2}$$

Obviously the equation had a root at $X = 0$, since at least one female was required to begin growth. Assuming that there would be a second root, corresponding to a level of stationary population, Lotka arbitrarily picked the simplest expression having two roots (two values of X at which $dX/dt = 0$). This left him with the approximation:

$$\frac{dX}{dt} = aX + bX^2. \tag{4.3}$$

Simple integration of this equation would have given him the logistic curve, but Lotka instead applied a different method of solving for X which corresponded to his general method of analysis in more complicated cases. Without repeating the steps of his solution here, it is sufficient to note that the result was the logistic curve.

Reviewing the various applications of the logistic equation by Pearl and Reed and others, Lotka remained unimpressed by Robertson's autocatalytic hypothesis. He suggested that the reason individuals followed the same growth curve as populations was that they consisted essentially of "populations" of cells. He proposed the term "autocatakinetic" in place of "autocatalytic" to describe this type of growth. This term was first used by Wolfgang Ostwald, who in 1908 published a growth study which independently duplicated the results of Robertson's articles of the same year.[31] Ostwald preferred the word "autocatakinetic" because it carried no implication of the underlying mechanisms of growth, and Lotka preserved his term for the same reason.

Like Pearl, Lotka referred to the logistic as a law of growth, where law was narrowly understood to mean an empirical relation between events, having no apparent connection to principles of a more general nature.[32] By this criterion, any other equation fitting the observations would have qualified equally as a law. But Pearl had come to regard his curve as lawlike in a much wider sense, possessing universal applicability. Unfortunately, in extending his theory he only enmeshed himself more deeply in a tangle of inadequate interpretation. Lotka avoided these problems precisely because he saw the logistic curve as the empirical statement it was, and he was therefore able to make more intelligent use of it. He perceived that an empirical law of this sort imposed limits in two ways. First, because the fundamental principles underlying the curve were unknown, the exact form of the equation had to be determined anew for each example. Second, it was not possible to extrapolate much beyond the observed events, because unknown factors might come into play outside the observed range and cause departures from the "law."

But a statement of this sort could still be useful as an entry into further analysis, as Lotka explained: "An empirical formula is therefore not so much the solution of a problem as the challenge to such solution. It is a point of interrogation, an animated question mark."[33] These comments were delivered in 1925 at the same meeting in New York at which Pearl and Reed had spoken and which Pearl had enthusiastically declared a success for the logistic curve.

Lotka developed his ideas around an analysis of the statistic r while employed at the Metropolitan Life Insurance Company. Before "interrogating" the logistic equation, he rewrote it as follows:

$$r = r_0 \left(1 - \frac{N}{N_\infty}\right), \tag{4.4}$$

where r is the rate of increase per head when the population was N (or dN/Ndt); r_0 is the maximum, or incipient, rate of increase (the same as r in equation [3.5]); and N_∞ is the limiting population (the same as K in equation [3.5]). The equation therefore redefined the rate of increase per individual as a function of numbers. It stated that the rate of increase at any time was given by the maximum or incipient rate of increase, diminished by a term proportional to the existing population, which was expressed as a fraction of the upper limit. Lotka interpreted the term $r_0 N/N_\infty$ as representing a complex of factors contributing to the decrease of the rate of growth; as such it was sensitive to changes in economic and social conditions, and was therefore difficult to analyze. The upper limit N_∞ or K was therefore of an economic character primarily, but it would also have a biological component in that for every value of N_∞ there would correspond a definite lower limit of fertility which would ensure that the population remained stable at that level.[34]

The parameter r_0, on the other hand, was amenable to more detailed analysis. In 1925 Lotka co-authored with Louis I. Dublin an article, "On the True Rate of Natural Increase," in which they showed that the conventional method of measuring the rate of increase as birthrate minus deathrate was misleading as a result of the effects of age distribution on these measures.[35] The intrinsic rate of increase was actually independent of the age distribution of the population. As a sequel to this study, Lotka probed the meaning of the incipient rate of increase in the logistic equation. With the American population as his test case, he calculated the maximum fertility of women in the late eighteenth century, at a time when the rate of growth was close to the maximum level. Expressing the fertility in terms of the average interval between births, he found that in the most fertile age group the interval was 13¼ months. Therefore the biological significance of the constant r_0, the maximum rate of increase, was that it was determined by the reproductive capability of the species in question; in this case, the physiological necessity in humans of allowing just over a year between successive births.

Later, Lotka fleshed out his study by a thorough analysis of the demographic characteristics of a population growing according to the logistic curve. A short talk on the subject was given to the American Statistical Association in 1931,[36] followed by a more detailed version presented to the International Union for the Scientific Investigation of Population Problems. The longer version was published both in the *Proceedings* of the meeting and in Pearl's journal, *Human Biology*.[37] A comparison of the theoretical characteristics with those actually observed in the U.S. population showed close coincidence, thereby offering confirmation that the population was in fact growing logistically, at least up to 1930.

Lotka always maintained a conservative attitude toward the logistic curve, regarding it as an approximation to actual trends. His analysis was aimed simply at discovering how much could be said about a population, given the reasonable assumption of logistic growth. By his suggestion that the logistic curve might best be seen as an "animated question mark," Lotka handed to Pearl a good argument for promoting the equation, while circumventing the criticisms which Pearl had invited with his rigid insistence that he had discovered a law of growth. But Pearl would have had to recognize that the value of his equation lay in the fact that it could be so easily derived from first principles, and that its constants *r* and *K* were biologically meaningful. The equation represented, in other words, an argument about the population which could be useful as a tool of research.

In this respect, the logistic curve was similar to the Hardy-Weinberg law in population genetics, which establishes the constancy of gene and genotype frequencies in a randomly breeding population. This statement had been called a law not so much from any confidence in its universality, but because R. C. Punnett wanted to honor his friend G. H. Hardy, who had quickly come up with this solution to a puzzle Punnett had brought to him.[38] Both the logistic curve and the Hardy-Weinberg law are deduced from a number of assumptions which can be falsified. Both greatly simplify the behavior of real populations. But by looking at how a population departs from the law, one may get a more realistic idea of the actual mechanisms underlying population growth or evolution. Knowing how a population deviates from the law tells one how to refine the initial assumptions to get a more accurate understanding of how populations behave.

Pearl stubbornly resisted letting go of his claim that he had discovered the law of population. Instead of making his equation meaningful by analyzing its assumptions, his explanations had a quality of overabstraction, which critics found hard to grasp. By treating the population as a discrete entity, like an organism, he abstracted it from its environmental context, and then "explained" its growth in terms of another abstraction, the logistic curve. As Pearl said of someone else, "Proving that one abstraction is wholly equivalent to another abstraction butters only metaphysical parsnips."[39]

The debate was not over yet, however. Pearl continued his crusade, seemingly undamaged and undaunted by the growing horde of infidels which beset him, as he settled into his institute and looked forward to at least five years of freedom to do as he liked. But there was one critic who was becoming rather insistent that he be heard. This was Edwin Bidwell Wilson, a physicist trained under Josiah Willard Gibbs, who had moved into the field of vital statistics in 1923 and settled at Harvard. Wilson's research interests began at this stage to overlap with Pearl's. As he observed Pearl, he began to develop an intense dislike for Pearl's flamboyant style. Wilson was the sort of man in whom such feelings simmer to a bitter and concentrated sauce. He possessed a wit to match Pearl's, and a good deal less gentlemanly restraint.

Wilson's Onslaught

Pearl had come to know Wilson when both were involved in editorial work for the *Proceedings* of the National Academy of Sciences. Over the years he developed a high regard for Wilson's abilities, little suspecting that the feeling was not reciprocated. The first hint of Wilson's dissatisfaction with Pearl's style came in 1923, when in the course of an article on the use of statistics in science, he made passing reference of a mildly critical nature to Pearl's *Biology of Death*.[40] The following February he delivered one of the DeLamar Lectures at Pearl's home ground, the School of Hygiene and Public Health at Johns Hopkins, publishing it later in *Science*. I do not know whether Pearl was in the audience or to what extent the published version represents what Wilson actually said in 1924. But if it does reflect his talk accurately, then the audience at Johns Hopkins would have heard him make a most peculiar argument, full of intense emotion, against those people "who seem for some reason to believe that a mathematical formula is eternally true":

> Their attitude is Shamanistic. They go through with magic propitiatory rites, idolatrous of mathematics, ignorant of what it can and cannot do for them. And I am not quite sure that the high priests of this pure and undefiled science do not somewhat aid and abet the idolatory.[41]

The "high priests" were Karl Pearson and George Udny Yule, both supportive of Pearl's work. Wilson was especially annoyed at Yule's exposition of the logistic theory at the British Association meeting in Toronto in 1924. In response, he devoted another article in *Science* to a satirical attack on Pearl's curve.[42] Using Canadian census data, he fitted them to an equation of the logistic form, but where the constants had been selected to indicate an accelerated, not a retarded rate of growth. In other words, Wilson's form of the equation predicted that "on some day apparently in the year 2020 the Canadian population will become infinite. *Dies irae, dies illa*!" Wilson tried again with a generalized logistic equation and found that the results "remained disconcerting." But, he added, "I am not a good curve fitter." Wilson openly admitted that it was "pushing the formula pretty hard" to distort it in this fashion, but he felt strongly that the logistic curve had no underlying theoretical justification and was, moreover, inadequate to describe a rapidly growing population like the Canadian.

He reserved his serious criticism for a second article devoted to Pearl and Reed's pamphlet on the future population of New York City.[43] They had estimated that New York and the surrounding area would reach a population of about 35 million. Working with Harvard astronomer W. J. Luyten, Wilson calculated the future populations for the three surrounding states of New York, New Jersey, and Connecticut by means of a logistic curve. They obtained a value of 22 million, which was less than that predicted for New York City alone. Therefore, they argued, the logistic theory led to the impossible result that the projected population of the whole was less than that of the parts. Pearl and Reed easily answered this point by claiming that Wilson had improperly fitted his logistic curve.[44]

Using a different method, they obtained a saturation value of about 50 million for the three states together, well above the New York City prediction.

Wilson had also based his argument on what he perceived to be a logical fallacy: the logistic theory could never be applied separately to the whole and its parts because two logistic curves when added together did not give a logistic curve. Therefore, at any time the sum of the component logistic curves would be out of line with the result predicted by a curve independently calculated for the whole area. He was sufficiently pleased by this argument to send it to Pearl's supporter, G. Udny Yule, and to A. L. Bowley, "a statistician of the hard-headed English type."[45] Bowley in turn referred it to Karl Pearson. The response, as communicated to Pearl by his friend, Major Greenwood, was one of derision toward Wilson:

> What is the matter with Wilson. I am almost sure you told me long ago that he was a very first rate man; will I am thinking he is a small minded ass and this opinion is shared by Yule and Karl the Great. He has sent separately to Bowley and Yule (I have not had it) an apparently unpublished paper demonstrating at great length that your population curve is non-additive. Bowley . . . was so delighted that he offered to send the great work to Karl which he did, and the great man, with the lines around his mouth hardened into the risus sardonicus, subsequently informed me that he wondered why a grown man should spend so many pages of algebra upon an obvious consequence and still more why he should suppose it invalidated any conclusions you had reached.[46]

Bowley also raised the criticism of nonadditivity at the 1925 meeting of the Royal Statistical Society, but without making much of it. Udny Yule answered him, pointing out that the logarithmic (Malthusian) law was also nonadditive, but that in any event the objection applied only if one wanted to use the logistic curve for predictive purposes in both the parts and the whole. As he had stated in his own article, he was not concerned with prediction: he reiterated that he saw no difficulty in applying the logistic equation to a large area where migration was subject to constraint, even when the formula did not hold for parts of that area.[47] Wilson, not to be put off by such arguments, announced that he had had special graph paper made up, on which the logistic curve could be plotted as a straight line.[48] He offered the paper at small cost to the members of the National Academy of Sciences, presumably in order that they might discover the deficiencies of the curve for themselves.

Pearl commented that Wilson's behavior stemmed from "a defense reaction arising from an inferiority complex based upon a lack of real understanding of biometric work."[49] But he clearly felt the sting of the criticism, for unlike Yule, he did place value on the predictive ability of the logistic curve. Rather than back down on this point, Pearl and Reed wrote a rebuttal to Wilson in which they showed that if the curves to be summed had the same rate of growth and were synchronous in time, their addition would give a logistic curve.[50] This was the condition that applied to the three states used in Wilson's paper, they felt. Otherwise the sum would not be exactly logistic but some other curve. Unable to disregard this possibility, they appealed to the generalized form of their equation to represent the summation. As the generalized equation could hardly be called a law of growth, this appeal did nothing to weaken Wilson's criticisms.

Up to this time Wilson had been but a minor irritation to Pearl. But he soon had the opportunity to inflict serious damage, and it was Pearl who handed him the ammunition. In the late 1920s, Pearl had turned to the statistical analysis of cancer incidence and was excited by its prospects. After analyzing hundreds of autopsy records, he found what appeared to be a negative correlation between cancer and tuberculosis. People with active tuberculosis seemed less likely to have malignant tumors than those either without tuberculosis or those with old, healed lesions. On this basis he set up an experimental program to see whether cancer could be treated with injections of tuberculin. After some preliminary work on rats conducted by a research associate, Pearl thought it desirable to extend the experiment to humans. He arranged for a cooperative project with the Harvard Medical School and St. Elizabeth's Hospital in Washington to test the tuberculin treatment.

Although the tests had barely begun, Pearl felt optimistic. After only a year, he knew he could not make any positive claims about the value of the treatment, but he did feel that the lives of a few of his patients had been prolonged because of it. His experiments had involved only a handful of people, all in the terminal stages of cancer, but Pearl felt that if the treatment offered any hope to patients, it was too important to hold back the facts. He published the preliminary results in *The Lancet* in 1929, with the conclusion that they justified at least continuing the research.[51] But he had no idea of how, or why, such treatment might work.

By this time the five-year grant for his Institute for Biological Research was nearing its end. The university was willing to apply to the Rockefeller Foundation on Pearl's behalf to continue the institute's work, and Pearl submitted a new grant application.[52] Before the proposal had time to be considered, however, a more splendid opportunity presented itself. His old friend, William Morton Wheeler, was about to retire as head of the Bussey Institution at Harvard, and Harvard had plans to reorganize its biological departments to develop a field of human biology. Pearl's name came up as the successor to Wheeler and as coordinator of the planned reorganization. Not only was Wheeler himself strongly in favor of Pearl, but he had the support of two influential men, Lawrence Henderson and Thomas

Barbour, who were able to convince President Lowell that Pearl was the right choice.

By coincidence, Lowell's wife developed cancer at this time, and there was talk of giving her the tuberculin treatment.[53] How her illness may have affected Lowell's decision is unknown, but in the spring of 1929 he offered Pearl the job. Pearl accepted.[54] The plans to continue his institute came to a halt, with the university withdrawing its application from the Rockefeller Foundation. Lowell's offer was unofficial, however; it still had to be approved by the Harvard Board of Overseers.

Hearing that Pearl was to come to Harvard, Wilson spun into action. Pearl had made some clear mistakes in the analysis of his cancer data and had reasoned from them much too quickly. Wilson was himself involved in the statistical analysis of cancer and was in a good position to discredit Pearl's research. Over the course of the next few months he waged a vigorous campaign to prevent Pearl's coming to Harvard. By mid-summer his criticisms of Pearl's statistical methods had attracted enough attention to put serious doubts in the minds of some of the overseers. Wheeler reported to Pearl the depressing news that Wilson had "stirred up the whole medical school, the department of economics, and at least four other departments to protest against your coming to Harvard."[55] He was of the opinion that Wilson, as well as some of the other opponents, were competing for the appointment themselves. Pearl retained a dignified silence throughout these trying months, relying on the lobbying efforts of Henderson, Barbour, and Wheeler to salvage his reputation. Although they were able to martial much support on Pearl's behalf in the form of testimonials from other scientists,[56] their efforts were to no avail. The overseers rejected Pearl's nomination in September 1929 by a vote of ten to nine.[57]

Pearl, weakened but not crushed, had anticipated this outcome. He told the president of Johns Hopkins that he wished to remain there. President Ames made discrete inquiries among the faculty: the feeling was that, although Pearl had made some mistakes, he was a valuable and stimulating person to have around, and that he should certainly stay.[58] But the Harvard scandal had created an awkward situation as far as the future of the institute was concerned.

The Rockefeller Foundation was hesitant to commit itself to support the institute to the extent Pearl had requested. Pearl's most important ally, E. R. Embree, was no longer at the foundation to argue on his behalf. Max Mason, the new temporary head of the Division of the Natural Sciences, visited Pearl's laboratory and was disappointed by what he found.[59] The work of the institute did not, in his view, appear to justify the money that had been lavished upon it in the previous five years. Moreover, the other biology departments at Johns Hopkins were expressing their needs for funds as well. Herbert Spencer Jennings, Pearl's old teacher and now Director of the Zoology Department at the Homewood campus of the university, thought that the institute's existence had hindered the other biology departments from getting outside support.[60] His own department had for years been severely short of space, with people forced to work in the greenhouse, the museum storage room, and the biology library.[61] No sooner had Pearl given up plans for the continuation of his institute, than Jennings made his own application to the Rockefeller Foundation.

The problem of funding was resolved to everyone's satisfaction when the university made a blanket proposal to the Rockefeller Foundation to cover the departments of zoology, botany, and plant physiology, as well as Pearl's laboratory. In 1930 the Rockefeller Foundation appropriated $387,000 for the support of biological research over a ten-year period, starting with $50,000 and decreasing by $2,500 each year, the decrease to be made up by the university.[62] The Institute for Biological Research ceased to exist, and Pearl was transferred back to the School of Hygiene and Public Health as the head of a new biology department consisting solely of his laboratory. There would be no question of Pearl enjoying quite the same degree of autonomy as he had with his institute. But if his reputation had suffered a blow, he was at least not badly off financially, although receiving less than before.[63] The other biology departments were at first pleased by the new arrangement, but some ill-feeling developed later when it was realized that Pearl was receiving a disproportionate share of the Rockefeller money and that the other departments were bearing the brunt of the annual decreases in the grants.[64]

Having succeeded in his campaign to keep Pearl from Harvard, Wilson had not yet fully spent his venom. In 1932 he published an article analyzing the statistical results of Pearl and others working on cancer and tuberculin, although without any reference to the value of the tuberculin treatment itself.[65] He concluded that the data did not warrant the inference that cancer and tuberculosis were negatively correlated. At the same time he launched an attack on what he perceived to be the indiscriminate use of statistics in science. Without mentioning Pearl, there is little doubt that he had Pearl in mind when he disparaged the work of certain unnamed disciples of Galton and Pearson as doing little more than childish whittling, accumulating shavings but creating nothing:

> They use the words and the data of science without appreciating what science is; their continuing boyish activities have become a bore, and the clutter they make has become a nuisance. . . . Certainly there seems to be something almost malignant about the multiplication of statistical detail, whether in material or in method, which so infiltrates and overgrows sound thinking as to bring about its early demise.[66]

Nor did the logistic curve escape one final barrage. In a lengthy and technical article published in 1933, Wilson expanded the criticisms he had made in his earlier articles.[67] Once again he unfairly included in the category of "logistic curve" equations having impossible values for constants, and where growth was accelerated to infinite values rather than retarded as the sigmoid curve would require. After detailed analysis of the simple logistic curve, he concluded that, although it might be useful for fitting census data, it could not be interpreted as a rational law of growth, nor could it be made the basis for forecasts. In short, after a decade of argument, he returned to Verhulst's conclusions of 1845, that until more observations were assembled the law of population remained unknown.

These conclusions had already been made several times by others, for Pearl's oft-cited "plain facts of experience" had long been in the line of fire. One critic voiced a complaint about the use of successive cycles of curves; provided that one divided it into small enough fragments, every continuous growing function could be represented by a good approximation of the logistic curve.[68] Demographers studying the growth of Chicago found that, in this region at least, growth had not followed a logistic path.[69] James Gray, zoologist at Cambridge, demonstrated how easily the same set of observations could fit two different sigmoid curves, leaving no way of choosing between them on the basis of fit.[70] Lancelot Hogben pointed out that the logistic curve, by its lack of uniqueness, lost all claim to being a universal biological law of population.[71] Sewall Wright seized on the same issue: any flexible mathematical formula resulting in a sigmoid shape could be made to fit the data.[72]

Following all these valid criticisms, Wilson's unremarkable conclusions sat rather limply, especially when juxtaposed against the intense rhetoric of the debate, with its flights into illogic on both sides as emotions took control over reason. But Pearl and Wilson were not merely arguing about whether the logistic curve was a law of growth. They were also arguing about money, prestige, and power. Their clash of styles, of methods and philosophies, of personalities, became critical only when their overlapping interests put them in competition with each other. The idealistic conventions of scientific discourse do not allow one to argue about money and power openly. Consequently, these concerns become transmuted into scientific controversies, and the struggle for temporal advantage unfolds in the intellectual arena. Although such debates often do involve incompatible philosophical commitments, which may hinder any real agreement on principle, it is also the case that they may perfectly well be decided on the basis of the facts at hand. Even Pearl knew enough to have realized that his claims for the logistic equation were unjustified. When such debates are not decided "on the facts," but are prolonged with a bitter and exaggerated vehemence, it is often because there are complex underlying material concerns involved. These may not be openly recognized, but they influence the logic and rhetoric of scientific discourse nonetheless.

The Demography of Animals

Pearl was a man of decisive temperament who could well be accused of slighting his critics, but to dwell on the polemical aspects of the logistic curve debate would be to underestimate his positive influence. Though he raised a chorus of criticism from Wilson and others, his activities did help to generate interest in experimental population biology. By continually emphasizing the links between human and animal populations, and between statistical analysis and experimental method, Pearl showed biologists that there were interesting questions to be asked of populations, involving a level of mathematics which need not be intimidating. As a result of his championing the methods of comparative demography, he was able to draw people into population biology who might not have entered it without his example. In generating controversy, he also generated positive discussion.

The furor over the logistic curve had raised many questions about curve fitting and the value of mathematical arguments as "animated question marks." Though Pearl did not address these questions himself, his associates on the sidelines of the controversy did. One of his researchers, Charles P. Winsor, who was to become an eloquent spokesman for the use of mathematics in biology, first became interested in the logistic curve when he was working as an engineer for the New England Telephone and Telegraph Company. At Pearl's suggestion he left his job there to join the institute's staff in 1927. As his small contribution to the debate in progress, Winsor compared the logistic curve to several other curves in use. The most suitable alternative was a sigmoid curve invented in 1825 by Benjamin Gompertz and used subsequently in various growth studies.[73] Winsor showed that both the logistic and Gompertz curves had similar properties for the empirical representation of growth, neither having a substantial advantage over the other.[74] He noted, though, that the differential form of the logistic curve could be more readily deduced by mathematical reasoning than the Gompertz curve. This advantage would ensure the logistic equation's continued survival.

Other institute workers added to the accumulating literature on the logistic curve and the demographic analysis of animal populations.[75] These studies did little to extend the theory of the logistic curve, but they served to establish examples of its use in the literature and so, through repetition, made its appearance more acceptable. The duration-of-life studies on *Drosophila* were influential because they demonstrated how actuarial methods might be applied to animal populations, and they suggested areas of research for others to follow up in the laboratory. The suggestion that density effects needed more study stimulated other biologists to make their own investigations on density and population growth.[76]

That Pearl's approach was not really ecological can be seen in a comparison with the methods of Royal Norton Chapman, who in his Minnesota laboratory was studying the population ecology of the flour beetle at this time. As an ecologist, Chapman saw the problem of population growth in terms of the interaction between the population and the environment. He was looking for the equivalent of Ohm's law in ecology. This law stated that the current of electricity at any point over a given time depended on the potential difference in the conductor and on the resistance offered to the current. Transferring the analogy to his beetle populations, Chapman thought that the density of a population depended on its "biotic potential" or maximum capability for growth, and the "environmental resistance" which acted to inhibit growth. The meaning of biotic potential was imprecise and included in some cases what should have been environmental resistance. Recognizing the imprecision, he tried inventing the concept of a "partial potential" to stand for the biotic potential of a species under a given set of circumstances, but this addition merely lessened the usefulness of the idea of biotic potential as he had first outlined it.[77]

Though his eagerness to embrace as a "law" of ecology an empirical relationship based on a false analogy with physics strikes us as naive, it reflected a common concern at the time to find a simple rule of thumb which would allow for some measure of prediction in ecology. The ideal, though not realized, was to be able to plug a value into an equation and to calculate how a population would respond to some environmental change. The purpose of laboratory study was to find the numerical values for those biotic constants and environmental resistances which could be plugged into the equations.

Pearl's experimental studies in animal demography were pursued with a different object from Chapman's. He was working on what was essentially an extension of the old heredity versus environment debate, trying to establish an order of priority for the different factors which affected the statistical features of populations. He wanted to know, for instance, whether the shape of the survivorship curve of a population was determined by the environment of the population or by its biological (genetic) makeup. Pearl therefore preferred to examine the influence of the population on itself by looking at behavioral and genetic variables which Chapman had not considered. Nor was Pearl interested in finding numerical values for the constants in his equations; he would examine the overall shape of his curves, seeing how they changed under different experimental conditions or in different genetic strains. This was a broadly comparative approach intended to show that "the biological principles underlying population phenomena are fundamentally identical in man and animals."[78] Of course, Chapman and Pearl were really studying aspects of the same problem. It was impossible truly to separate environmental and biotic components in the experiments, for the population created its own environmental conditions to which it then responded. But their methods and goals differed according to their different interests.

Pearl and Chapman's differences may have produced a mild rivalry between the two laboratories, for the remarks of one of Pearl's students, E. Cuyler Hammond, hinted at some friction between them. In one article, Chapman, without referring to Pearl, had argued for the use of beetle populations as opposed to fruit flies, because he felt that the genetic variability of the flies raised doubts about the validity of making any exact calculations of the biotic constants.[79] Hammond reacted rather testily to this criticism in a later article, indicating it had been a prickly issue in Pearl's laboratory. "Pearl and his associates have never attached any special significance to the arithmetic values of particular points picked from their fitted curves, which values might be called 'constants' by Chapman's definition," he wrote, adding: "They were not interested in selling life insurance to flies or beetles either, but in showing sequences of events which when general and repeatable can be considered as biological principles, and this they have done."[80]

Although this was a novel method for ecology, it was not simply the novelty but also the force of Pearl's personality which made people respond to it. Friedrich Simon Bodenheimer, a young entomologist from Jerusalem, stopped in Baltimore to see Pearl in 1931 on his way to visit Chapman in Minneapolis. He was afterward most impressed by Pearl's will and energy.[81] In his 1938 book *Problems of Animal Ecology*, he included a supportive account of Pearl's fruit-fly studies.[82] He was equally enthusiastic about the logistic curve, to the extent that he tried to interpret certain demographic, cultural, and economic phenomena of human history as consequences of the population's position on the curve.[83]

If an ecologist wanted to learn about statistics and demography in the 1930s, Pearl's laboratory was a good place to start, for Pearl's broadly intellectual view of biology could still be inspiring, whatever the errors of reasoning his critics had pointed out. Pearl enjoyed working with ecologists, especially when they were of the caliber of Thomas Park, who arrived as a postdoctoral fellow in 1933 and stayed on as an instructor until 1937. He had become interested in population ecology while studying under W. C. Allee in Chicago. In response to Chapman's results with flour beetles, he decided to investigate these animals on his own. During his years in Pearl's laboratory, he studied the effect of environmental factors on the growth of beetle populations, and especially the beetles' effects on their own population through the alteration of the surrounding flour. His results showed, as distinct from Pearl's own explanation of the density effect by direct interference, that such alteration of the flour had an important influence on fecundity.[84] After he returned to Chicago in 1937, he expanded his study to include competition between two species and developed over the years an active program in laboratory population ecology, centered on the problem of competition.

Pearl's activities generated enough interest and controversy that as late as the 1940s biologists were still testing and revising some of his conclusions. F. W. Robertson and J. H. Sang, working in Lancelot Hogben's department, analyzed the density effect that Pearl had suggested was the *vera causa* behind his logistic law.[85] They discovered that changes in the quality and quantity of food were the most important factors influencing the rate of egg laying in *Drosophila*. The interference effect that Pearl postulated seemed to be due to competition for food: if the flies were adequately fed, crowding produced only a slight decrease in fecundity. Their results did not invalidate Pearl's hypothesis of interference effects due to crowding. They did suggest, however, that his use of this argument to support the logistic curve hypothesis had no force. Bodenheimer later tested their results and agreed with their conclusions: "There can therefore be no doubt that the logistic curve of growth is a purely formalistic rule. Not much remains of the regulating density-factor which we all accepted thirty years ago as the explanation of the logistic curve of population growth."[86] These words were published in 1958: Pearl's influence was remarkably long-lived. The logistic curve itself never died: it became and remains a useful starting place for the mathematical treatment of population dynamics. Part of what made it acceptable in ecology was the fact that it was a convenient expression in which the constants had biological significance. The other part was its high visibility: it had become solidly established in the literature within a decade of its introduction. We shall run into it again.

With this sketch of some of the questions which spurred the growth of laboratory population ecology in the 1930s, I have run ahead of my story. At the same time that the logistic curve and Pearl's ideas were being noticed by the ecological community, the theoretical study of population ecology was expanding on several fronts. By the 1930s a variety of strategies in population analysis was on display, but it was not obvious how they might be integrated with field practice. The questions, conflicts, and confusions which were so visible in the logistic curve debate permeated all areas of population biology that involved the use of theory in these decades. The debates over facts, principles, and strategies, with signs of competitive struggle for limited resources, are ones that we shall find reiterated constantly in the chapters that follow.

5

Modeling Nature

As more attention was paid to population dynamics in the 1920s, the one point on which there was general agreement was that this was a subject of considerable confusion and obscurity. Raymond Pearl had taken drastic steps to reduce the confusion in the biology of population growth with his bold announcement that he had discovered the law which would bring the subject to order. The storm that followed, while it caused some to distrust the crude simplicity of mathematical arguments, encouraged others to recognize the insights that seeking simplicity might offer.

Pearl's was not the only mathematical line of attack to appear during the twenties. In other areas as well, population events were being investigated with mathematical tools: these ranged from the primitive implements of the biologist to the more elaborate tools of the expert mathematician. The different approaches to population analysis reflected not only different levels of expertise, but also different perceptions of the purpose of the inquiry. On the one hand, the economic biologist wanted to predict and to control specific populations; on the other hand, the mathematician was interested in creating a general theory of the struggle for existence as an imaginative exercise. A variety of strategies sprang up within a few years of each other. These met with a mixed but polite reception at first, but then they began to foment controversy as ecologists confronted the implications of allowing mathematical thinking in this empirical discipline. In a curious twist to the story, one of the earliest advocates of mathematics in ecology, W. R. Thompson, turned out to be one of its most vigorous opponents by the mid-thirties. Before we can understand why Thompson changed his mind, it is necessary to review what these different mathematical offerings were. In this chapter, I shall discuss the main strategies which appeared in the 1920s and early 1930s; in chapter six I shall discuss their reception and the controversies they engendered.

Hosts, Parasites, and Mathematicians

The ability of mathematics to suggest conclusions not possible through observation alone had been noted early in the century, though without attracting much attention. Sir Ronald Ross, winner of the Nobel Prize in 1902 for his work on the cause of malaria, was prompted to take up a mathematical argument when he found that, even after the *Anopheles* mosquito was known to transmit the disease, there remained much popular resistance to the idea that the best way to control malaria was by controlling mosquito populations.[1] The prevailing view in the field was that the incidence of disease was not closely correlated with the numbers of mosquitoes in an area. This crude impression was summoned as an argument against the insect control measures that Ross had advocated.

Hoping to reconcile these observations with his understanding of the disease, Ross tried using a mathematical description of the relation between mosquitoes, malaria, and humans. His analysis showed that the disease would not maintain itself unless the proportion of mosquitoes was at a certain level, and that above this level a small increase in mosquitoes would cause a large increase in the incidence of malaria. Here was a possible explanation of the apparent lack of correlation between the two populations.

Ross characterized his approach as the "*a priori* method," meaning that he began by making assumptions about the cause of an epidemic; constructed a set of differential equations to describe the situation based on these assumptions; deduced the logical consequences of the mathematical argument; then tested these theoretical results by comparing them with observations. The use of the term "a priori," now commonplace for this type of modeling, can be misleading, for it suggests also that the model is constructed prior to experience, which Ross certainly did not intend. All of the theoretical treatments I shall discuss in this chapter are arguments of this type, proceeding logically from cause to effect, though their trains of reasoning are very different. Ross called his method the "Theory of Happenings," a general title intended to suggest the wide applicability of the method, not only to the quantitative study of epidemics, but also to "questions connected with statistics, demography, public health, the theory of evolution, and even commerce, politics, and statesmanship."[2] Although he began using mathematics as early as 1899, his theory of happenings appeared in 1911 as an addendum to his book *The Prevention of Malaria*. His method differed from the more usual a posteriori approach, which began with the observations; fitted analytical laws to them; and worked backward to the underlying causes. This was the method commonly used in statistics. In epidemiology it had recently been applied by John Brownlee, who built upon the researches of the nineteenth-century statistician William Farr.[3]

Whereas in epidemiology the statistics were available to support both methods of reasoning, in ecological studies the same wealth of information about life histories and populations was lacking. By the 1920s, at least one ecologist had become impatient for its accumulation. William Robin Thompson, a Canadian entomologist working for the U.S. Bureau of

Entomology, was rapidly coming to the conclusion that much order might be thrown into this confused subject by reasoning, as Ross had recommended, from a set of assumptions to their logical effects, even in the absence of a systematic body of observations.

Thompson had been hired in 1919 to do research on biological control at the bureau's European Parasite Laboratory in France.[4] He was already well known at the bureau from his student days when he had worked at the Gypsy Moth Parasite Laboratory in Massachusetts, starting in 1908. Afterward the bureau had sent him to Cornell University for graduate work, then to Italy to study the alfalfa weevil with Filippo Silvestri, one of Europe's leading entomologists. In 1913 he resigned to pursue his biological studies at Cambridge and Paris; this was followed by a stint in the Royal Navy Medical Service during the war. In 1918 he returned to Paris and was shortly rehired by L. O. Howard to study the biological control of the corn borer, a European insect that had recently become a pest in Massachusetts.

At the European laboratory, Thompson was in charge of analyzing the relationship between the corn borer and the parasites that controlled its abundance in its native habitat. In thinking about this problem, it occurred to him that a mathematical approach might be fruitful as a way of suggesting new hypotheses. The use of mathematics to disentangle the causes that together produced a given effect had impressed him after reading D'Arcy Wentworth Thompson's new work, *On Growth and Form*, which had appeared in 1917.[5] Although these volumes had nothing to do with ecology, they showed how an understanding of mathematical relationships could be brought to bear upon a variety of problems related to the growth and structure of plants and animals. In problems of growth and form, mathematical laws could be applied with confidence because they were based upon the physical laws governing the organic and inorganic worlds. Reading these studies, Thompson carried this reasoning into his ecological problems. If ecological interactions were found to display an underlying regularity, and if this regularity could be described mathematically, then mathematics might serve as a theoretical basis for population ecology.

Thompson found the evidence of this regularity in the work of his two mentors, L. O. Howard in America and Paul Marchal in France. Both had conducted field studies of insect populations in the 1890s and had independently made the same observation: that parasite and host populations seemed to fluctuate together in definite cycles.[6] This empirical evidence gave Thompson all the excuse he needed to explore the mathematics of the case: "For he who says periodicity, regularity, rhythm, says the possibility of a mathematical representation."[7] Starting in 1920, he began to use simple algebraical expressions to describe the relations between parasites and their hosts. The idea was to find an expression for the number of hosts in each generation, taking into account the various factors which would be most important in determining the growth rate of the population (i.e., reproductive power of host and parasite, proportion of sexes in each species, and number of parasite eggs laid in each host). Assuming different initial values for the host and parasite populations, Thompson used his equations to calculate the change in the numbers of each species in successive generations and the change in the percentage of parasitism in each generation.

His equations showed that, in the early stages, the presence of a parasite did not appreciably prevent the host from increasing. The host population increased much as it would had the parasite been absent, becoming more of a nuisance with each generation. At some stage, though, the parasite population would rapidly outstrip the host and would bring about a sudden crash to extinction in the host population within a single generation. Thompson realized that this prediction was not quite right: in reality, such extinctions did not occur. Rather, the host population would merely be reduced to a low level, where it would no longer be a nuisance. But his theoretical results seemed to support observations Paul Marchal had made in the field, that changes in the numbers of insects and their parasites sometimes followed a pattern of large oscillations, each having a slow ascending period and an abrupt downward descent, with the parasite apparently causing the decline of the host.

Thompson was confident that his equations, though they simplified the biology of the interaction, did express the basic relations between host and parasite. Moreover, they offered hope for the success of future biological control programs. His findings suggested that biological control might not show any effect for several generations, but that when its effect was finally achieved, its results would be more complete and of longer duration than would be possible with mechanical or chemical means of control. He argued that, important as it was to know the details of particular cases, real progress in entomology could only be made by uncovering the general laws expressing the process underlying each particular case. Once those laws were encapsulated in formulas, it would be possible to examine particular cases and to draw conclusions of value in practical work:

> Not that these conclusions will always be rigorously in accord with the facts. Far from it. But one can at least consider them as a theme on which nature embroiders infinite variations of reality and by virtue of this they constitute a theoretical base for our work.[8]

Fifteen years later, Thompson would retract these words and confess the error of his youthful ways, but for now he plunged enthusiastically into mathematical ecology, calling upon the advice of more expert mathematicians when the problems exceeded his own abilities.

His colleagues proved to be harder to convert to mathematics than he had hoped. The officer in charge of the corn-borer research, to whom Thompson first showed his results, decided that this work was too mathematical for most entomologists. He advised Thompson that he would have trouble publishing such work in American entomological journals. Instead, Thompson wrote up his results in French and published them with Marchal's help in the *Comptes Rendus* of the Academy of Sciences in Paris.[9] Another manuscript, sent to Cambridge for publication there, came back with criticisms based not on the mathematics but on entomological problems.[10] His conversations with other entomologists revealed that their hesitation to adopt his methods was not because of the mathematics, but because his equations required biological information about the insects that was unknown; specifically, knowledge of the effective rates of reproduction of the populations in the field. His colleagues could argue from strength that it was too early yet for mathematics, that what they needed was more biological research, especially more research on the life histories of the animals under study. Thompson was sympathetic to these charges and set to work to modify his equations so that they could be used by the practical biologist.

If most of Thompson's colleagues were sluggish in responding to his mathematics, there was one mathematician who was quick to grasp the relevance of these studies to his own grand schemes. Alfred Lotka was just contemplating writing his book when Thompson's articles appeared. He had progressed steadily in working out his general method of systems analysis, with special attention to two-species interactions, and was always on the watch for concrete examples to illustrate the method's usefulness. Keeping a close eye on the literature, he found his examples, first in Ross's analysis of malaria, then in Thompson's entomological writings.

Lotka seized upon Ross's research as soon as it appeared in 1911 and incorporated the malaria example into his general study on evolution.[11] He saw in the malaria case an opportunity for a thorough study which would both illustrate his method and indicate how it might be applied: it became the focus of an exhaustive mathematical treatment, some of it written with Frank R. Sharpe, a mathematician he had met at Cornell several years before. Together they published several detailed analyses of the Ross equations in the 1920s.[12] Lotka's work sprang directly from Ross's but was refined and expanded to deal with additional problems missing from Ross's treatment; in particular, the development of the course of a malaria epidemic in its entirety (as opposed to a consideration of the equilibrium condition only), and a study of the effect of a time lag caused by a period of incubation of the malaria parasite.

These promising beginnings in mathematical epidemiology engendered few disciples. Ross expressed surprise in 1915 that so little mathematical research should have been done on such an important subject, especially given the quantity of statistics that had accumulated by then. In Edinburgh, W. O. Kermack, a biological chemist, teamed up with A. G. McKendrick, who had been interested in the mathematics of growth since the 1910s, to extend Ross's analysis in a series of papers published in the 1920s and 1930s.[13] But Lotka's efforts did not seem to impress Ross greatly, despite his desire to encourage further research in this area. Lotka sent him a copy of his book for a review in *Science Progress.* Ross had it reviewed anonymously by "an expert biometrician, who has spoken very favourably of it".[14] The review made note of all the places where Lotka had used the Ross equations, but otherwise found the theories to be "somewhat ultra-speculative."[15] As for the series on malaria that had preceded the book, Lotka had hoped that it would help him to make a connection with the Rockefeller Institute, at the time a center of experimental epidemiology, but his efforts earned him nothing in the medical field. In general, the practice of epidemiology was developing largely along laboratory, experimental lines on which the mathematical approaches of Ross and Lotka made no impact.[16]

Though the case study into which Lotka had poured so much effort finally came to nought, there were other areas where his method could be applied. When Thompson's articles appeared, Lotka accordingly brought the host-parasite example into his analysis as well. He did not fail to notice that Thompson's fluctuations provided the empirical support for the rhythmic oscillations that Herbert Spencer had deduced as the necessary outcome of the balance between the forces of increase and decrease in populations. Lotka's method differed from Thompson's in that he used a continuous-time scale, rather than the discrete-time scale with the generation as the unit of time which Thompson had used. The change gave Lotka greater flexibility but was less realistic for insect populations.

Lotka's analytical method began by describing the interactions between species as a set of simultaneous differential equations. His technique was based on methods used for the mathematical description of the dynamics of chemical reactions. For the detailed analysis of systems of differential equations, he was indebted to the researches of Henri Poincaré and Charles Emile Picard. The procedure Lotka used was an early version of what later came to be called "general system theory," which was developed by Ludwig von Bertalanffy after the Second World War. General system theory as Bertalanffy conceived it was both a point of view and a method.[17] The object was to develop mathematical techniques of analysis which could be used to model the interrelationships between the component parts comprising any sort of system. The word "system" was therefore interpreted very broadly, much more broadly than Lotka had done, so that, as a mathematical technique, systems analysis could be applied to fields as different as biology, information theory, economics, or sociology:

the idea was to make the technique applicable no matter what the particular features of the systems were themselves. Lotka did not imagine his technique as having such broad uses, but the mathematical procedure outlined in his discussion of population dynamics was identical to the one that Bertalanffy later used.

The method itself followed from Lotka's "fundamental equation of kinetics," where the increase in a given component of the system, X_i, is expressed as a function of all the other components, X_1 to X_n, and of all the environmental parameters, P, and the genetic parameters, Q:

$$\frac{dX_i}{dt} = F_i(X_1, X_2, \ldots, X_n;\ P, Q). \qquad (5.1)$$

The components, X_n, could refer, for example, to the masses of a group of species living in a biological association. To simplify the problem, the first thing he did was to assume that both the environment and the genetic constitution of the species were constant: in this way, the P's and Q's could be neglected. The fundamental equation was then applied to each component separately, X_1 to X_n, and the form of the function F (which was unknown) was approximated by means of a Taylor series expansion. This produced a series of differential equations which linked the increase of each component X_i to every other like component in the system. The next step was to solve the equations (using a mathematical technique derived largely from Henri Poincaré) and to examine the roots of the solution. From this general examination, certain conclusions could be drawn about the behavior of the aggregates under scrutiny. The value of the method, as Lotka explained, was that fairly specific distinctions could be made between different cases without having complete information about the exact functions, F.[18] When applied to single populations, the method could be used to derive the logistic curve with the appropriate assumptions following the Taylor series expansion.

In the case of the host-parasite and predator-prey relationships, which Lotka tended to lump together, he used his method to depict the interaction between the two species with much greater generality than Thompson had been able to do. For instance, he showed that under certain conditions the interaction would give rise to continual oscillations of the two populations: this was the mathematical representation of the periodic fluctuations that Thompson had noted but had not fully analyzed. A more refined analysis showed that the interaction might also take the form of a damped oscillation, where the fluctuations gradually decreased in magnitude and approached a stationary level (Figure 5.1).[19]

From these mathematical models, Lotka was able to suggest some tentative conclusions. For example, under the original assumptions it would be impossible for one species to eliminate the other, although a

a

$$T = 2\pi/\sqrt{r_1 d_2}$$

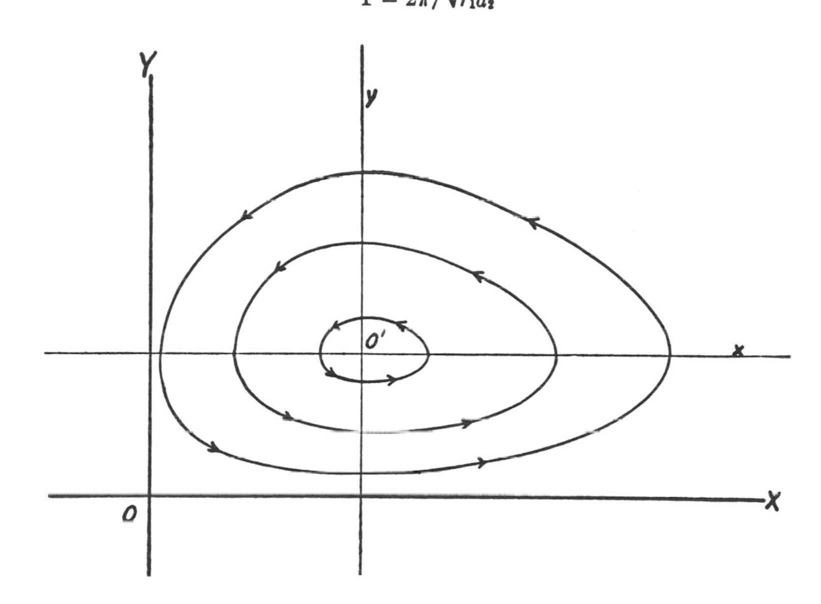

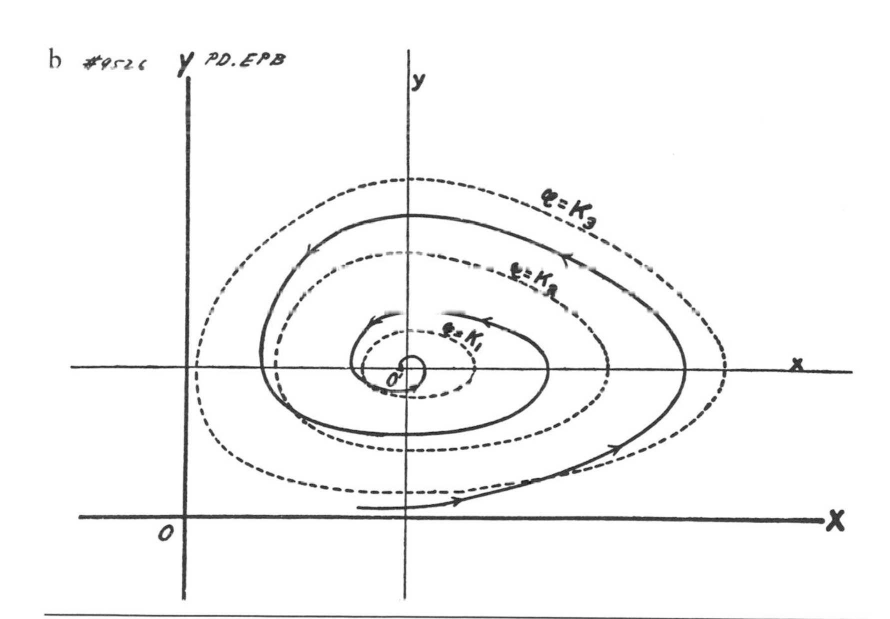

FIGURE 5.1. Curves illustrating oscillations in parasite and host populations. Each axis represents the number of one of the populations. (*a*) A cyclical process continuing indefinitely, the classic Lotka-Volterra oscillations; (*b*) Lotka's more exact treatment, resulting in a damped oscillation. (From A. J. Lotka, *Elements of Mathematical Biology*, New York: Dover Press, 1956, pp. 90, 91.)

predator might diminish its prey enough to make it vulnerable to other influences. Moreover, the addition of a second prey species would not necessarily benefit the first prey, but might to the contrary increase its chances of extinction, because the predator would no longer be constrained by the decrease of the first prey. Lotka thought this hypothetical case might have a counterpart in fisheries: heavy fishing of a common species could accidentally cause the extinction of a rarer species taken along with it, when normally the rare species would be protected by its scarcity.[20]

The predator-prey interaction and the more detailed malaria case study were but two examples used as illustrations for one part of the physical biology program. Different initial conditions could give rise to different models of interaction. Not all of these would correspond to situations in the real world, but Lotka illustrated them in his book along with the realistic models (Figure 5.2). At all times he was concerned not just to solve a specific example but to show how a general method of systems analysis could be used for any related problem. This was a fundamentally different approach from that followed by either Ross or Thompson, neither of whom specifically tried to place their studies in a broader context, mathematically or biologically.

But what was for Lotka an advantage, namely, the generality of his method, also stood as an obstacle to its acceptance, for it seemed too all-encompassing to be useful in applied research. Lotka was greatly disheartened to find that even those, such as Ross and Thompson, who might have been expected to appreciate the value of this work, failed to take much interest in it. We might speculate, then, about how he felt when he read in the pages of *Nature*, a year after the appearance of his book, that a celebrated Italian mathematician had come up with an almost identical solution to the predator-prey problem.

Lotka and Volterra

The mathematician in question was Vito Volterra, who held the Chair of Mathematical Physics in Rome and was already known for his work on the theory of elasticity and the theory of integral and integro-differential equations.[21] He had been interested in the idea of applying mathematics to the biological and social sciences as early as 1901, but only in 1925 did he turn abruptly and with remarkable perseverence to mathematical ecology. Indirectly he came to this line of work through his daughter, Luisa, who was an ecologist and was engaged at the time to a young marine biologist named Umberto D'Ancona. D'Ancona was engaged in an analysis of some market statistics of the Adriatic fisheries around the time of the First World War. He found it odd that there appeared an unusual increase in certain predaceous species during the war years, when fishing had almost

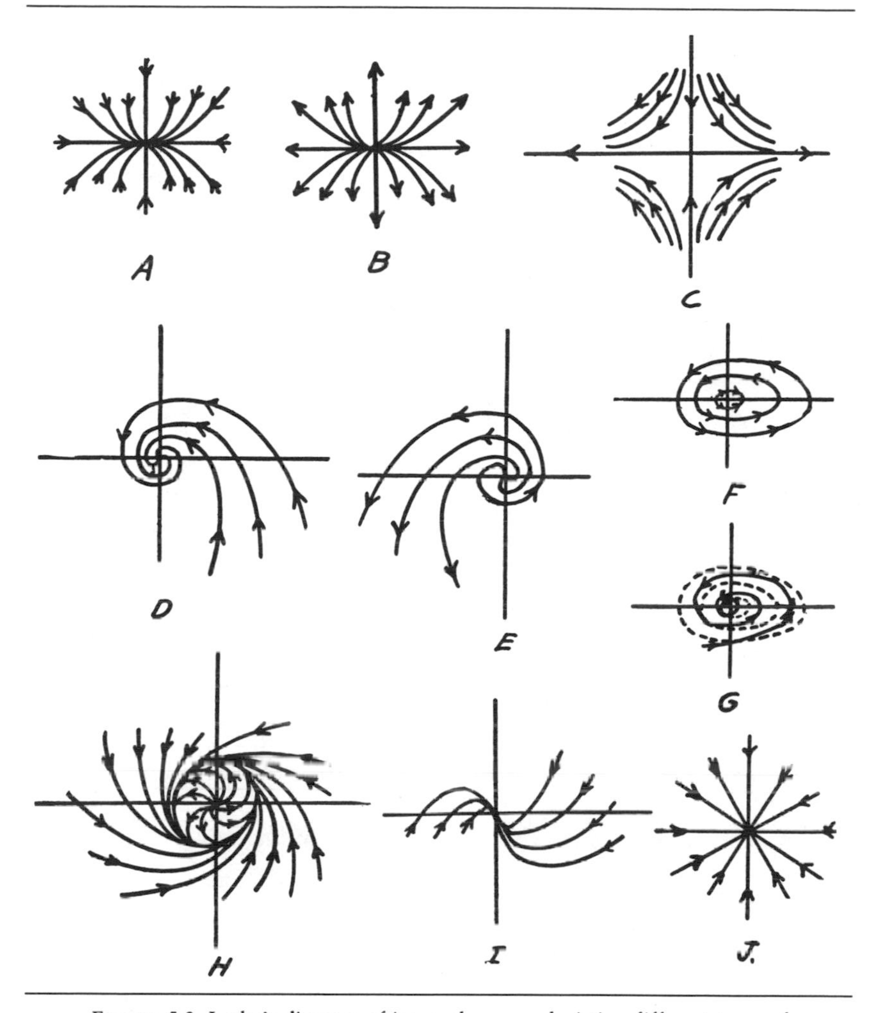

FIGURE 5.2. Lotka's diagram of integral curves, depicting different types of equilibrium, stable and unstable, in systems with two dependent variables. Cases F and G correspond to the host-parasite relations shown in Figure 5.1. The Ross malaria equations gave rise to two conditions, a stable equilibrium of type A and an unstable one of type C. The other types were included for purposes of illustration but were not associated with concrete examples. (From A. J. Lotka, *Elements of Mathematical Biology*, New York: Dover Press, 1956, p. 148.)

ceased. Pondering the problem with his future father-in-law, he wondered if there could be a mathematical explanation for these changes.

Volterra took up the problem in earnest in 1925, and for the remaining fifteen years of his life became absorbed in the ramifications of this question. By 1926 he had published an elementary account in Italian of the interactions of species in a biological association.[22] A short resumé of this article published in *Nature* in the same year brought his work before the English scientific community[23] and the ever-watchful eye of Lotka. This article was the merest hint of what was to come; a light theme which was to form the basis for ever more imaginative and elaborate variations.

The article began with an analysis of a simple ecological problem, the interaction between a predator and its prey population. From there he

VITO VOLTERRA, 1860–1940
Photograph courtesy of Brandeis University Library

moved to a general analysis of population interactions, both predatory and competitive, sketching rapidly the outlines of a broad theory which would ultimately, he hoped, lead to an understanding of the dynamical behavior of the entire ecological community. Volterra considered his analysis to be part of evolutionary biology, an attempt to investigate, along mathematical lines, the day-to-day interactions of organisms as a first step toward a fully mathematical, general theory of evolution. It was, in short, the mathematical theory of the struggle for existence.

It was in the discussion of the two-species, predator-prey interaction that Lotka's and Volterra's work overlapped. With the identical problem to Lotka's, that of the numerical relations of a predator to its prey, Volterra easily derived the same equations and conclusions that the interaction would give rise to periodic oscillations in the two populations. This was the same problem Lotka had discussed in 1920 and which had first attracted Raymond Pearl's attention to his work.

It is interesting to note that the general method, apart from the applications made by Ross in epidemiology, had also been used in military strategy analysis. Frederick William Lanchester had used a similar technique to analyze combat during the First World War, and Lewis Fry Richardson independently proposed the technique to analyze combat in 1919.[24] Lotka appears not to have known of these other applications, but he certainly was aware of the parallels between the use of models in a military context and his own methods of model building. As he suggested in the *Elements*: "It is well worth considering whether interesting light may not be thrown on various problems of biological conflict, by the use of models designed to imitate the biological warfare somewhat after the manner in which the war game imitates the armed conflict of nations."[25]

For the conflict between predator and prey, Lotka expressed the relationship verbally as follows:

$$\begin{bmatrix}\text{Change in}\\ \text{number of}\\ \text{prey, } N_1\text{, per}\\ \text{unit of time}\end{bmatrix} = \begin{bmatrix}\text{Natural}\\ \text{increase of}\\ \text{prey per unit}\\ \text{of time}\end{bmatrix} - \begin{bmatrix}\text{Destruction}\\ \text{of prey by}\\ \text{predators per}\\ \text{unit of time}\end{bmatrix}. \tag{5.2}$$

$$\begin{bmatrix}\text{Change in}\\ \text{number of}\\ \text{predators,}\\ N_2\text{, per unit}\\ \text{of time}\end{bmatrix} = \begin{bmatrix}\text{Increase in}\\ \text{predators per}\\ \text{time, as result}\\ \text{of ingestion}\\ \text{of prey}\end{bmatrix} - \begin{bmatrix}\text{Deaths of}\\ \text{predators per}\\ \text{unit of time}\end{bmatrix}. \tag{5.3}$$

These equations can be translated into differential equations for the prey species, N_1, and the predator, N_2:

$$\frac{dN_1}{dt} = r_1 N_1 - k_1 N_1 N_2, \tag{5.4}$$

$$\frac{dN_2}{dt} = k_2 N_1 N_2 - d_2 N_2, \tag{5.5}$$

where r_1 is the coefficient of increase for the prey species (births minus deaths); d_2 is the coefficient of mortality of the predator; and k_1 and k_2 are constants.

Comparing the growth of the prey species to that of the predator, we have

$$\frac{dN_1}{dN_2} = \frac{N_1}{N_2} \frac{(r_1 - k_1 N_2)}{(k_2 N_1 - d_2)}. \tag{5.6}$$

Integrating this equation and graphing the solution, a family of closed curves results, which depicts the continual oscillation of the two populations. Both Lotka and Volterra obtained the same cyclical solutions, although with slightly different reasoning: Lotka followed the reasoning of the above word-equations, whereas Volterra used what he called the "method of encounters."[26]

The method of encounters began with an analogy between physical and biological aggregations. Volterra likened the individuals in a biological association to molecules of a gas in a closed container. This comparison gave him his mathematical point of entry into the problem of how they interacted. In statistical mechanics, the number of collisions between particles of different gases is proportional to the product of their densities. In the same way, Volterra supposed that the events in a biological aggregate depended on the number of "encounters" between individuals, where the probability that one individual would encounter another would be proportional to the product of the numbers of both species. Each encounter was presumed to lead to an immediate result for each individual, which might be favorable, unfavorable, or neutral. The method of encounters was the mathematical counterpart of Lotka's method of systems analysis using Taylor series expansions, described above. It allowed Volterra to set up equations describing the course of each species and to generalize further to the case of *n* species.

As this summary indicates, there were a great many simplifying assumptions involved in both cases. In particular, the equations did not allow for the influence of the density of the population on its own rate of increase; that is, the populations always tended to increase exponentially and not logistically. The populations were also assumed to be homogeneous: each individual the same age and size as every other and invariable over time. Finally, each encounter between predator and prey would have an immediate effect on the individuals involved. The oscillatory behavior therefore presumed the simplest kind of interaction between the two populations, excluding other biological and environmental variables.

Despite recognition of these unrealistic simplifications, Volterra expressed his conclusions in the form of three "laws." The first, the "law of the periodic cycle," stated that the fluctuations of the two species were periodic in nature and depended only on the initial conditions and the various coefficients of increase and decrease. The second, the "law of conservation of the averages," provided for the constancy of the average numbers of the two species, all else being constant, no matter what their initial numbers were. The third, "the law of the disturbance of the averages," stated that, if an attempt were made to destroy the individuals of the two species of predator and prey uniformly and in proportion to their numbers, the average number of the prey species would increase and that of the predator would decrease. This last law would imply, for instance, that a temporary halt in fishing would benefit the predator, a prediction which seemed to be borne out by D'Ancona's independent observations based on the statistics from the Italian markets. The apparent confirmation of theory by observation suggested to Volterra that he was on the right track.

When Volterra's article appeared in 1926, Lotka had settled into his job at the Metropolitan Life Insurance Company and was pursuing his demographic studies with energy and satisfaction. He had little time left over to devote to physical biology, but Volterra's piece stirred him to write a letter to the editor of *Nature*, pointing out areas of overlap with his book.[27] The letter was published with a reply from Volterra acknowledging Lotka's priority in certain areas, but indicating quite rightly that there were still important differences between them. From this polite exchange there began a brief but mutually respectful correspondence between the two, from which Lotka drew a welcome measure of moral support:

> Your very kind interest and good wishes are of material assistance to me in renewing my energies on a topic in which there has not always been much encouragement for my work and in which I had almost come to feel that I would not be able to do much more hereafter, but I feel differently since reading your letter.[28]

For all Lotka's satisfaction that a mathematician of Volterra's rank should have indirectly endorsed his results, the problem of priority worried him. He decided to write a review article discussing the relation between his and Volterra's contributions. Raymond Pearl promised him space for it in *The Quarterly Review of Biology*. The manuscript began with a justification of the use of a chemical viewpoint in biology; dwelt at length on the competition and predation cases as discussed by Volterra, with supplemental analysis by Lotka; moved to a consideration of energy relationships; and finally ended with a philosophical discussion of perception and consciousness and their relation to physics.[29] For the most part, the intended article was a condensed version of the *Elements*, with more detail on the areas pertinent to Volterra's investigations. As Lotka worked on the manuscript, Volterra himself was busy elaborating and publicizing his own work: he expanded the early treatment with a detailed study published in 1927 in Italian, while shorter articles and translations of the 1926 paper appeared in French, English, and Russian.[30]

Slowly Lotka's discussion took shape. By mid-year in 1928 he had decided to publish some of the technical details ahead of time, for reasons he explained to Pearl: "The fact is that I am somewhat in fear of anticipation of my other work at the hands of Volterra. So far he has barely touched on the phase of the matter which I am taking up, but there is always a risk that he might branch out in that direction."[31] Apart from this short piece, hidden in a mathematical journal, nothing of more general interest appeared for three years. In 1931 Lotka had written only one short article on the mathematical theory of capture,[32] similar to his earlier mathematical paper. Beginning with the predator-prey equations, he considered the conditions under which a predator found and captured its prey in a given territory. This was an extension of the discussion in the *Elements* in which he had used the chess-game analogy, and it resembled in its details and tone a problem in military strategy. The article, published in 1932, was to be followed by two others, on "frequency of capture" and "influence on inter-species equilibrium of modification in the characteristics of competing species."[33] These were never published, but a companion article on Volterra's competition equations, also published in 1932, may have been a preliminary version of the third part.[34] In the meantime, Pearl was still awaiting the review article for his journal.

In 1931 the appearance of a book by Volterra, *Leçons sur la théorie mathématique de la lutte pour la vie*,[35] increased Lotka's sense of urgency to publish. The book was compiled from a lecture series Volterra had given in the winter of 1928–1929 at the new Institut Henri Poincaré in Paris. Here Volterra elaborated some of the ideas he had introduced in the earlier papers, with new refinements to make up for the lack of realism of the early models. He had early on made a distinction between two types of biological associations, conservative and dissipative ones. Conservative systems were analogous to frictionless systems in mechanics: in a biologically conservative system the oscillations set up by the interactions of the species remained constant, with none of the species going extinct or increasing indefinitely in a finite time. This was the situation represented by the predator-prey oscillation described above.

But Volterra believed that absolutely conservative systems were ideal cases, which only approximated the natural situation. It was more likely that natural associations were dissipative, that is, the fluctuations of the species were damped and the association tended toward an equilibrium state, analogous to the effect of internal friction in material systems. The main difference between the two systems was that the dissipative system took account of the effects of a population's size on its own growth (as, for instance, if the populations grew logistically, rather than exponentially). These effects would tend to dampen the oscillations between the different species. The distinction therefore introduced a more realistic modification to the idealized system of continual oscillations.

The *Leçons* developed the mathematical distinctions between these two systems and included a chapter discussing time lags as well. Volterra referred to these time-lag effects as "hereditary phenomena"; heredity referring not to descent, but to the fact that a population's history could influence its present behavior. This meant that an encounter need not have an immediate effect, but might be noticeable only after a certain time interval; the analogy here was with conditions of retardation or drag in mechanics.

Although these modifications were meant to make the models more realistic, the book was essentially an elaborate mathematical argument, based on the principles of mechanics as they might be applied to biological aggregations. Even in their more sophisticated form, the models were based on many unrealistic assumptions, from which some rather far-reaching conclusions had been deduced. But as Volterra fully admitted, this was to be seen as a work in pure mathematics, even if it was couched in biological language. It was, as he wrote, the rational phase of the study of biological associations. Those who would embark on the applied phase would require more profound discussion, based on fact and experience, of the initial hypotheses.

Although Lotka found the discussion in the book admirable, he was disappointed to find his own work given superficial treatment in the historical chapter, written by D'Ancona, concluding it. His grievance was directed not so much at Volterra, who had conscientiously mentioned Lotka's work in all his publications after 1926. Rather, Lotka was concerned that other writers, influenced by Volterra's brief references, were perpetuating the impression that Lotka's work was insignificant in comparison with Volterra's. His feelings were aggravated by the appearance of several publications which favored Volterra's work and seemed to be influenced by personal connections to Volterra himself.

One of these was a review article on mathematical biology published in 1927 by Joseph Pérès, a mathematician who had helped W. R. Thompson a few years earlier.[36] Pérès was also a former student of Volterra and a later collaborator, so it was not surprising that his article reflected Volterra's contributions overwhelmingly. Lotka was dismayed when he came across the article a year later, as he wrote to Pearl:

> He gives eleven pages to Volterra's work and three or four footnotes to mine. I think you will agree with me that this is definitely a case of displacement of the center of gravity. After reading his paper, which is very good as

far as it goes, I was more pleased than ever at the opportunity which you are kindly offering me of coming out before American Readers with a statement of the situation as viewed from our side of the issue.[37]

Each discussion of mathematical ecology that appeared seemed to compound the injury Lotka felt. Karl Friederichs, a German entomologist, included a discussion of the researches of Thompson, Lotka, and Volterra in his zoological text published in 1930.[38] All Lotka saw was that he had given Volterra's work much greater prominence. Feeling by now wholly exasperated, yet remaining strangely silent despite his acute sense of being neglected, Lotka explained his actions to Friederichs:

Perhaps I might add that when I allowed the correspondence relating to the Volterra matter to close after a very brief letter from me, this was not because I acquiesced in the position that Volterra had taken, but because I have a very strong personal aversion to priority disputes. . . . Through the kindness of the editors of one of our journals I have been given an opportunity to express myself at length on the matter and I hope to do so in due course; but as I have already stated, my time is greatly occupied and I prefer to give my efforts to productive work rather than to squabble about priority. Nevertheless, the occasion seems to call for some action on my part if I can possibly get down to it.[39]

Lotka never did fulfill his plans to publish an assessment of Volterra's work in relation to his own. Neither the review article for Pearl, nor a review of Volterra's book which he had intended to write for Pearl, were ever completed. Just at that time, however, a new opportunity to express his ideas was offered to him through the mediation of a Russian geophysicist and mathematician, Vladimir Aleksandrovich Kostitzin. Lotka made use of the offer, not to debate with Volterra, but to summarize his physical-biology point of view for a different audience and to gather his latest results in demography into one book.

Kostitzin, originally trained as a geophysicist, left Russia for Paris in the late 1920s, whereupon he came into contact with Volterra, who was giving a lecture series at the Sorbonne, and developed an interest in mathematical biology. He maintained close ties with Volterra throughout the 1930s.[40] In 1933 he was working on a mathematical study of symbiosis and parasitism (his wife was a parasitologist), which was to be published as part of a series on biometry and statistical biology edited by the French geneticist Georges Teissier.[41] Kostitzin's own work had been influenced by Lotka's extensions of Thompson's results, and in connection with this series he contacted Lotka to see if he would be interested in contributing to it. Lotka immediately suggested a two-part treatment of biological aggregations, the first to be devoted to demographic phenomena in a single population, the second treating of mixed populations comprising several species, the whole to be tied to the principle of "evolution."[42]

The final product, published as *Théorie analytique des associations biologiques*,[43] was somewhat narrower in scope. The first part, which appeared in 1934, was an overview of the issues and general methods described in the *Elements*. One noteworthy change was Lotka's more explicit appeal to ecologists, with references to the relevant writings of R. N. Chapman and Charles Elton. His discussion of Volterra was minimal, confined to a single footnote correcting a historical error of priority in the case of *n* species interacting. The second part did not appear until 1939 and was entirely demographic, focusing on human populations. His intended treatment of mixed populations of several species never appeared, except in so far as it was included in the general discussion of the first part. But Volterra and D'Ancona had subsequently published a monograph on biological associations in 1935 as part of the same series,[44], and it is possible that Lotka (or the editors) felt that his treatment of the same problem would be redundant.

Lotka's worry that the promotion of Volterra's work was threatening his own position was excessive. The mild priority dispute, which ensured Lotka at least a footnote, however brief, in Volterra's articles, helped to disseminate news of Lotka's research to the proper audience. The tendency to lump Lotka and Volterra together, however, also helped to obscure the differences between them. On the whole, Lotka's emphasis on energy relationships and the economic tone of his writings continued to be overlooked by biologists. But in the case of predator-prey interactions, at least, his priority was firmly established, and the equations with oscillatory solutions describing the changes in the two populations came to be known as the Lotka-Volterra equations.

By the late 1930s the differences which stemmed from Lotka and Volterra's different ideas of physical biology became more apparent. Lotka had become totally involved in demography (although he did keep abreast of the biological literature relevant to population studies), and this emphasis came to dominate his later work. The English revision of the *Théorie analytique*, to be called *Analytic Demography*,[45] was to be even more restricted to human populations, with analysis of certain demographic problems which he had not considered in the French version. He died in 1949 before completing this work. Whereas Lotka was focusing ever more closely on the case of single-species populations, trying to extract as much specific information as was possible through demographic analysis, Volterra was moving toward a more sweeping statement of the principles of mathematical biology. This involved the more conspicuous use of physical analogy, to the extent that he defined mathematically a quantity called "demographic energy" (actual and potential), which was conserved in the same way that the energy of physics was conserved.[46] He also defined "demographic work" and the "principle of least vital action," all of which was a direct transfer of the methods and concepts of physics into biology. These were merely more exaggerated uses of the same analogical reasoning which Volterra had employed in his early articles.

Concurrently with these highly mathematical studies, which had by the late 1930s covered enough paper to be widely known, if imperfectly understood, in the ecological community, there arose yet another mathematical strategy to add to the range of choices available to ecologists. This one appeared independently from the far side of the globe—Australia. It bore some resemblance to Thompson's and Volterra's approaches, and some important differences. It grew from a student's rude remark, and before the decade was out, it had lit the fire of one of the most heated controversies in population ecology.

The Balance of Nature

While Thompson, Lotka, and Volterra were pursuing their different courses in mathematical ecology, an entomologist working in Australia, Alexander John Nicholson, was beginning to work through his own puzzled thoughts on population regulation. He had been lecturing in entomology since 1921 at the University of Sydney, his first academic position following his studies in zoology, chemistry, and botany at the University of Birmingham, England, where he received his B.Sc. and M.Sc. degrees.[47] At Sydney he found little time for research, for he had first to organize a whole new subdepartment of entomology. At the start there was neither equipment nor material suitable for teaching, and he had to spend much of his time collecting and photographing insects for class use.

One of the rewards of teaching surely comes when students express skepticism about the established truths they are handed in the classroom, for it is by such rude queries that teachers are sometimes jolted to reconsider familiar arguments. So it was with Nicholson. He had taught that one of the means by which populations were limited was through the limitation of the food supply. One of his students, having answered an examination question as he had been taught, finished by asserting that, nevertheless, he did not believe it: the worst pests did not consume all of their food supply, even without artificial control measures. This observation was not new to Nicholson, but it caused him to think more carefully about how populations were controlled in nature.

He reasoned that an increase in a population of insects would bring about an increase in its enemies as well, which would prey more heavily on the first population. A species and its enemies would therefore tend to reach a balance at which the number of prey was just sufficient to support as many predators as would destroy the surplus number of prey produced. This idea resembled the earlier arguments of Stephen A. Forbes, especially his suggestion that species and their enemies would tend to develop a "common interest" which would produce a balance in nature. Nicholson then concluded that a population may, in a sense, be thought to limit *itself*, because it would induce greater opposition to further multiplication as it grew. From these ideas he developed a theory of population regulation based on the importance of competition within a species as the main regulating mechanism.

His first opportunity to apply these ideas came in 1927, when he was faced with the duty of writing an address as the retiring president of the Zoological Society of New South Wales. He had collected a large number of slides illustrating mimicry and protective coloration, and he decided to give the address on that topic. This talk soon grew into a larger study of mimicry which, along with other manuscripts dealing with population regulation, became his thesis for the Doctor of Science degree.[48]

By 1930 he had expanded this work into a massive, but still largely speculative, account of the possible mechanisms of population regulation. His efforts to publish it as a book came to an abrupt end when the referee turned it down. He picked what he thought were the most salvageable sections for separate publication: these dealt with the host-parasite interaction, which had been developed more precisely than other examples in the manuscript.

Nicholson's initial argument was nonmathematical, but like a mathematical argument it was deductive in nature, based on a set of simplifying assumptions, to which were added a few arithmetical computations. The computations were not based on field data, but were hypothetical numerical examples that helped him to depict his argument graphically in the absence of exact information. The hypothetical examples led to unexpected conclusions. His reasoning suggested that the interactions between host and parasite populations would lead to a system of oscillations that increased over time. That is, as a population began to swing back to its equilibrium position, it would tend to go too far, producing an unstable situation of ever-increasing oscillations.

Nicholson was disconcerted by this result. He appealed to a physicist colleague at the university, Victor Albert Bailey, for some mathematical help. Bailey converted Nicholson's verbal arguments into mathematical form and came up with the same conclusions, although he was able to state them with greater precision. Together they worked out more details of the argument, considering various initial assumptions and showing how different conclusions could be derived. The basic theory was summarized in two articles published in 1933 and 1935: the first, written by Nicholson, gave the verbal and arithmetical argument, while the second covered much the same ground with the addition of Bailey's mathematical proofs.[49]

Bailey's point of view reflected that of a physicist, just as Volterra's had. He considered the movement of parasites in search of hosts to be analogous to Maxwell's theory of the mean free path of a particle in a gas. In keeping with the comparison to the dynamic theory of gases, he assumed that density was uniform, and that search proceeded randomly in the population as a whole. Because Nicholson had based his argument on discrete-time intervals—that is, assuming a definite succession of generations (as opposed to the continuous-time models of Lotka and Volterra)—Bailey's mathematical verifications also used the same method. On his own, however, he extended the study to continuous interaction and worked out some further mathematical details in separate articles.[50]

Quite apart from Bailey's use of physical analogies to construct the

ALEXANDER JOHN NICHOLSON, 1895–1969
Photograph courtesy of Commonwealth Scientific and Industrial Research Organization Archives, Australia

basis of the mathematical argument, Nicholson's conception of the problem made use of an analogy which depended on the idea that there existed in nature a balance. This was based on the observation that population densities changed in response to changes in the environment. To convey his ideas, he employed the image of a population functioning like an instrument or a machine. The balance was conceived to be analogous to that of a balloon floating in the atmosphere. As the ambient temperature changed from day to night, the balloon would undergo changes in height and volume, continually rising and falling as its position of equilibrium with the surrounding air moved. In the same way, Nicholson thought that population densities were continually tending toward a stable level in relation to fluctuating environmental conditions. Experimental studies of the kind carried out by Pearl, Chapman, and others seemed to support the impression of balance gained from field observation; after a period of growth, a laboratory population would attain a certain steady density which represented its position of balance in the laboratory environment.

But the fact that a population would tend to stabilize at a given density under given conditions did not imply that populations were actually controlled by external conditions, such as climate. If, as Nicholson argued, the existence of balance implied the existence of a controlling factor, then it was also the case that such a controlling factor had to be responsive to changes within the population itself. Climatic effects usually were felt irrespective of the density of the population. A true control had to act with increasing severity as the population density increased. Nicholson felt that there was only one factor which met this requirement of density-dependent action, and this was competition. Competition by its very nature became more severe as density increased; it therefore had to be the mechanism behind population regulation. Organisms could not be thought of as having direct and immediate rapport with the environment at all times, rather they were indirectly responsive through the mediating influence of competitive relations with members of the same species. This was not to deny that sudden climatic changes did at times kill off portions of the population, but only to assert that such effects were not responsible for the balance of populations. The existence of competition was therefore inseparable from the idea of regulation.

In another metaphor, Nicholson compared the controlling function of competition to that of the governor on a steam engine.[51] Just as the governor responded to the weight of different loads on the engine by adjusting the steam output and thereby varying the power, in the same way a change in environmental stress caused the level of competition to rise or fall, until the density was again adjusted to balance the stress. Competition was not as sensitive as the governor of an engine, however, with the result that fluctuations in density would occur as the balance was readjusted. These metaphors did not always clarify Nicholson's reasoning. Competition was a vague and broadly defined concept. Without a clear connection to practical work, this deductive method would strike many as excessively abstract. Concrete experiments which would help to put the ideas into context would follow only after a delay of several years.[52]

The purpose of his theory was not just to contribute to economic entomology, however, but to clarify the role of natural selection in relation to the balance of nature. In his earlier study of mimicry, he had been struck by the fact that well-camouflaged species seemed to be no more successful than their relatives that lacked this adaptive property. This observation suggested that the success of a species (as seen by its numbers) was somehow independent of its possession of a given adaptive trait. Natural selection was responsible for producing adaptive characteristics, such as mimicry, but it was not responsible for the success of the species.

To modern ears this argument sounds odd. Nicholson imagined that competition acted as a counterforce to natural selection in the fixing of genetic traits during the course of evolution. He argued as follows: when individuals with advantageous characteristics appeared, they would tend to be preserved by natural selection. But their preservation would cause a population increase as well, until competition was so intense that some of the members of the original population would be destroyed. Gradually the new, favored type would come to replace the original type. Natural selection (preservation of new types) was seen as a disturbing influence which disrupted the balance of nature; whereas competition restored and maintained balance during and after "selection," enabling new types to replace old ones.[53] This was a rather literal interpretation of natural selection. Some adaptations therefore were of little value to the species as a whole, because they arose completely by competition within the species. This conclusion seemed compatible with his studies of mimicry in butterflies, because the degree of mimetic resemblance appeared to reach a level of perfection far beyond its effect on the general viability of the species.

This argument threw a strange twist into Darwin's discussion of natural selection. For Darwin, the corollary to the struggle for existence was that the structure of an organism was related, often in subtle ways, to that of all others with which it had to compete, or on which it fed, or from which it had to escape. Both intra- and interspecific competition could determine given adaptive structures. But whereas for Darwin natural selection acted through competition as part of one process, Nicholson made a sharp distinction between natural selection as a disruptive mechanism and competition as a regulatory mechanism.

Nicholson was trying to explode what he took to be a common but wrong belief, that natural selection had two functions: to select and to produce balance among populations. He felt that this idea of natural selection was an unfortunate example of teleological thinking. He argued, to the contrary, that natural selection functioned only to select and not to produce balance; that is, it improved adaptation but had nothing to do with the regulation of populations. He did not say where this common view originated, but recalling Stephen Forbes's nineteenth-century merger of Darwinian natural selection with Spencerian arguments about balance, it is possible that the view to which Nicholson objected was the product of a similar combination of Darwin and Spencer that went into early ecological theory. Nicholson believed that the view that natural selection created a balance in nature was caused partly by an erroneous concept of adaptation as the fairly close adjustment of animal to environment. But Nicholson did not think that ecological studies really supported this view of adaptation. Once it was understood that adaptation did not imply a precise balance of organism and environment, he felt, it would follow that an improvement in adaptation would be seen as having nothing to do with the balance or limitation of populations. His argument is at times difficult to unravel, but it illustrates the complexity and diversity of opinion surrounding the interpretation of natural selection, adaptation, and population regulation in the decade leading up to the modern synthesis.

Nicholson and Bailey were interested mainly in animal populations. For practical purposes they narrowed their view of competition to include only that occurring when animals were engaged in a search for essential resources: for instance, a parasite in search of its host. Under these conditions, competition depended on two basic properties: first, the species' power of increase; second, the individual's ability to exploit the surrounding territory to gain what it needed for survival. The density at which competition would be felt would depend on these two properties. The second property included a wide category of specific traits, such as the efficiency of organisms at finding, capturing, and utilizing resources, as well as the efficiency of their prey at avoiding capture.

Their discussion reflected the same awareness that Lotka had shown of the need to consider the detailed behavior schedule of the individuals, interpreted in energetic terms, before species relationships could be properly understood. Lotka was hampered by lack of data and was not able to carry his analysis very far. Nicholson and Bailey experienced the same obstacle, but they tried to overcome it by gathering all these behavioral and energetic terms into a single measure. These were summed up in the characteristic which Nicholson called the "area of discovery." This was defined as the area effectively explored by an average parasite individual, but it was intended to represent all the things that affected the efficiency of animals as they searched for resources over a given territory in their lifetimes. Without knowing exactly what these efficiencies were, species could still be compared by measuring their respective areas of discovery.

The rest of the theory followed from a detailed consideration of this problem, analyzed under a variety of hypothetical circumstances which might be expected to approximate real situations.

Nicholson and Bailey first considered the types of parasitism under which a steady state would be produced. They then moved to an analysis of the situation where a population was removed from its equilibrium position and would tend to return to its steady state level. Interpreted as a problem of competition while searching for food, and based on discontinuous interaction, the host-parasite relation took on a fundamentally different aspect than that given it by Lotka and Volterra. The "Lotka-Volterra equations" did not take into consideration the effect of competition from members of the same species. Moreover, although Volterra had later considered the effects of time lags in the results of an encounter, neither had included the delays which would result from the age distribution of the populations. That is, they assumed that individuals were born mature. In Bailey's analysis, both sorts of delays were taken into account. He found that when the age distribution in particular was considered, he did not obtain the steady state oscillations of the Lotka-Volterra model, but rather the unstable system of increasing oscillations that Nicholson had first found so disconcerting.[54] Nicholson and Bailey knew that these increasing oscillations were not found in nature. They suggested that the result of these oscillations would be the breaking up of the population into many widely separated groups, each group waxing and waning, finally disappearing and being replaced by new groups. In the predator-prey interaction, Nicholson concluded that, although the same conditions necessary for oscillations existed, they would likely be less violent in nature and would tend to produce a stable system of oscillations, rather than the unstable one of the host-parasite interaction.

In general, Nicholson and Bailey hoped for a more exact treatment of the problem of population regulation than any of their predecessors, with more careful consideration of the alternative outcomes that would result from different biological assumptions. But apart from the differences in their orientation and in the nature of their specific conclusions, their results were of the same character as those of Lotka and Volterra. They found that the interactions between species would, all else being equal, produce oscillations in the two populations, as distinct from oscillations caused by external environmental conditions. These conclusions were not meant to be exact representations of nature, but to indicate the ideal behavior of populations under simplified conditions. As theoretical predictions, they were intended to serve as guides for experiment and observation.

Nicholson and Bailey planned to write a series of five articles on the subject, of which Nicholson's 1933 paper was a summary, but only the first part of the joint series was published. Bailey sent it to Lotka in 1933 in manuscript form, with a request for Lotka's help in finding a suitable publisher for the series.[55] Lotka was irritated to find that his own work was hardly discussed, but he complied with the request and sent it to Raymond Pearl, though coupled with a testy letter to Pearl expressing his annoyance at not receiving proper credit.[56] Pearl sided with Lotka and rejected the article, adding that it appeared far too speculative for his journal,[57] but more likely feeling that it was too mathematical. Lotka sent the manuscript back to Bailey with a long letter explaining his dissatisfaction that his priority and contributions to biology had been overlooked in recent literature.[58] The article finally appeared in 1935 in the *Proceedings* of the Zoological Society of London. It began with a paragraph summarizing Lotka's work, but made the point that Lotka's equations seemed too general to yield the specific kinds of conclusions which Nicholson and Bailey were after. Moreover, they had not been able to derive their theory from Lotka's fundamental equations. In general, they felt a greater affinity toward Volterra's methods, which mirrored Bailey's own image of the population in analogy with the theory of gases.

Differences and Similarities

The researches of Thompson, Lotka, Volterra, Nicholson, and Bailey represented the principal lines along which theoretical population ecology developed in the 1920s and 1930s. Each was guided by a different method of reasoning, reflecting the different backgrounds and different goals of the authors.

Thompson's strategy was by far the most cautious and the most realistic. He believed in formulating a problem strictly in biological terms first, using mathematics only to simplify the statement of the problem. Given the lack of biological information on the populations he studied, his analysis did not take him very far. Nevertheless, he began optimistically, trusting that the use of mathematical models would give entomology the predictive ability of the physical sciences and would in turn guide the applied strategies of economic entomologists.

Nicholson shared many of Thompson's goals for applied science, though he had broader interests in evolutionary biology as well. He was on the whole much more prone to speculation than was Thompson, for although he had not had time for much research, he had time to teach and to think. He had an imaginative, metaphorical perception of the population, one that would later give Thompson much cause for complaint. But with Bailey's help, he hoped to create models which were both precise and realistic, incorporating better assumptions based on what was known of the behavior of host and parasite populations. They perceived the increased realism of their models to be a major improvement over Volterra's models.[59]

Volterra, on the other hand, came to the mathematical theory of the struggle for existence from his background in classical mechanics. His ways of thinking, his ideas of science, were those of classical mechanics. "All of us in our generation," he wrote in 1907, "were raised with those principles that a modern world calls mechanicist; and indeed, that all phenomena, at least those under the domain of physics, could be reduced to phenomena of motion and could be brought within the orbit of classical mechanics, was a dogma adhered to by every school and whose origin is lost in the remote Cartesian philosophy."[60] In 1925, through a chance inquiry, he seized the opportunity to extend this worldview to biology; it was natural that he would try to create, in essence, a biological mechanics.

The success of mechanics was in turn due to the use of the techniques of calculus. A few starting hypotheses, though not very realistic, would allow the problem to be represented mathematically with calculus. By seeing how well the mathematical predictions accorded with reality, the initial hypotheses could be adjusted to make them more realistic. The method that Volterra used therefore began with generality and worked toward greater realism. It started with a coarse view of nature and by a series of steps approached the fine reasoning of the geometer.[61] But in order to create this first, coarse view, Volterra had drawn heavily upon analogies taken from physics and used as heuristic devices. He had let his imagination run. If the metaphors were too abstract from a biological point of view, he could calm the reader with the assurance that he was, after all, engaged in a work of pure mathematics.

It was in the use of analogies that his methods were in greatest conflict with Lotka's. Lotka had come from the same tradition in physics, but he developed his analogies differently. He saw physical biology as being based on identity of type between physical and biological systems: this led him directly to the study of matter and energy transformations. When Lotka spoke of "energy" it was in the same sense as that understood by physicists; his use of the term "dynamics" denoted the study of energy transfers through the biological system. For Volterra, "biological dynamics" meant the enunciation of energetic principles in biology analogous to the ones in physical dynamics, such as the conservation of energy and the principle of least action. He did not look at energy exchanges in the population, but at the transformation of a wholly metaphorical "demographic energy." Lotka's particularly careful habit of thought, his attention to the meaning of words, his precision in the use of analogies, and his skepticism of metaphorical entities taken as realities, reflected the training in science which he had received from John Henry Poynting. These were

Poynting's habits of reasoning, which had so inspired Lotka in his student days and had guided him in the development of his ideas.[62]

Lotka was interested in uncovering laws of nature, following the model of physics, but he did not adopt Volterra's course. He was aware that the equations which he and Volterra had developed independently were formal statements which need not have any deeper significance. He felt it necessary to go beyond formal expressions in order to deduce necessary relations from known principles. In this way, it might be possible to arrive at a law which was not merely an empirical rule, but "a law of nature that brooks no exception."[63]

Analysis of this kind had to be based on a realistic perception of the individual, which for Lotka meant treating it as an energy transformer capable of a wide range of adaptive strategies to ensure its survival. But his analysis was not conducted with reference to an *actual* individual, species, or population; rather, the individual itself was idealized to represent a general class of energy transformer, for which there might be many examples in nature. As Lotka explained:

> It will not be necessary or even desirable to deal primarily with specific living organisms, but with transformer *types* possessing properties characteristic of the physical *modus operandi* of living organisms. The kind of problem then to be studied will be the relation between the distribution of matter in the system on the one hand, and on the other the particular properties and variation in properties of the several types of transformers of which the system is composed.[64]

By idealizing the organism in this way, precision was lost but generality increased. A problem dealing with the interaction between a "pursued" transformer and a "pursuing" transformer, for example, could be reduced to a problem of geometry. Using mathematical techniques of analysis, one could then discuss, for instance, the influence of density and distribution of refuges in the territory on the probability that the "pursuer" would capture the "pursued." The hypothetical organism was a model in an analogous sense to the Carnot heat engine in physics. Carnot's engine, which existed only on paper, was composed of perfectly conducting and insulating parts through which heat was transferred and work performed. Though an ideal case, it illustrated the underlying physical reality later expressed in the second law of thermodynamics. In the same way, Lotka imagined that his models of energy transformers would lead to general principles, based on physical and biological reality, which would govern all transformers of that type.[65] An example of one such principle was his "law" that evolution proceeded in such direction as to maximize energy flow through the whole system.

The result of this point of view was that his predictions were qualitative and were framed as comparisons between types of situations. Such predictions were furthest removed from the goals of applied science; it is no wonder that they were the least used and appreciated at the time. Lotka came to recognize the inevitable gap in communication between himself and biologists, but he did not try to close it. In 1945 he still referred to his work as a special branch of physics and not of biology. But his demographic work fared better, as will be seen in chapter six.

Despite the radically different strategies represented in the glut of theories that emerged in these years, they all had the same underlying objective: to show that theoretical, mathematical approaches had a place in biology. Put even more strongly, they wanted to show that theory could *guide* experiment and research, and that it was not worth waiting until all the facts were in before engaging in speculation with the help of mathematical models. As Lotka wrote to Volterra, knowing he would have the latter's full support, "I believe that it is necessary for us to deliberately overcome a certain repugnance which one feels towards such extreme conventionalization and to proceed with the work in the hope that the first crude steps may turn out in time to have been necessary preliminaries for a more perfect treatment of the subject."[66]

In the next chapter I shall describe how biologists responded to these choices of style and these grand claims. There were treasures here to satisfy many wants, and to provoke many jealousies. The range of responses extended from enthusiastic acceptance to hostile rejection. Many steered a middle course and remained interested but aloof, adopting, as Bacon had remarked of an earlier age, the prudent mean between "the arrogance of dogmatism and the despair of skepticism."[67]

6

Skeptics and Converts

"Grant a Mathematician one little Principle, he immediately draws a consequence from it, to which you must necessarily assent; and from this consequence another, till he leads you so far (whether you will or no) that you have much ado to believe him." So Fontenelle in 1686; since then, little had changed. When ecologists got wind of these mathematical treasures, they too were constrained to believe these gems were genuine. The closer one delved into the mathematical details, the greater was the danger of losing sight of the real world. All this mathematics baffled them and made them feel vaguely apprehensive.

The fear of losing reality was felt especially by economic entomologists. One such skeptic was Harry Scott Smith, in charge of biological control research in California. He had worked in the Parasite Laboratory of the Bureau of Entomology in Massachusetts before taking charge of the parasite work for the state of California in 1913. In 1923 he was appointed associate professor at the University of California and worked at the Citrus Experiment Station at Riverside. His background and research interests were similar to W. R. Thompson's, but he remained unimpressed by Thompson's mathematical forays. What was needed, he thought, was more qualitative information about how environmental conditions restricted populations and more information on the life histories of the animals.[1] Economic entomology was molded by the realities of applied science; attitudes toward mathematics were always colored by the relevance of these studies to agricultural practice and especially to the controversial status of biological control. As for Volterra's work, Smith could make neither head nor tail of it because he knew no calculus.

Lack of mathematical training was a common problem. Thompson had run up against this obstacle in 1921 when the man in charge of his research judged his mathematics too difficult to be of use to entomologists. Thompson himself was no mathematician: he relied on other mathematicians to help him out of difficulties. He appreciated Volterra's work at the outset, not for its mathematical details, which he never discussed, but because it helped him to envisage the host-parasite relationship in a general way. In 1930 he declared that Volterra's were the most important analyses to date and that he found them useful in dealing with the practical problems of insect control.[2] His papers on biological control in the early 1930s helped to acquaint English-speaking audiences with Volterra's research and those portions of Lotka's work that were related to it. But on the whole, he remained reserved in the face of these elaborate mathematical studies.

Such reservations were reasonable, for only the best mathematicians could expect to approach these works with any real comprehension. It is therefore surprising to see, balanced against Thompson's judicious restraint, the unabashed enthusiasm of his American colleague Royal Chapman. Chapman could not possibly have understood what he was being enthusiastic about, but that did not slow him down one bit.

He had already spotted Thompson's research in the mid-twenties as the most exciting work to appear in entomology for some time. He took a year's leave of absence in 1926–1927 to develop his own ideas about population ecology and to visit the major entomological centers of Europe. He planned to continue his experiments on the ecology of the flour beetle, an animal easily handled in the laboratory. Stopping first at Rothamsted Experimental Station at Harpenden, England, he worked out his experimental design under the guidance of R. A. Fisher. From there he headed for Thompson's laboratory at Hyères, France, where he arrived in early 1927. Back in Minnesota, his course on insect ecology was being taught by an entomologist from Montana, William C. Cook, who coincidentally chose *Elements of Physical Biology* as one of the texts. He wrote to Lotka that he found his book a "constant source of inspiration."[3]

Chapman meanwhile had polished off his beetle experiments and had discovered Volterra's work, a likely source being Pérès's popular account of 1927. What he had discovered was not just a new analytical tool but something of potentially greater value: a symbol of status. Mathematics, he believed, would raise the lowly status of ecology to the dignified level of the physical sciences. He believed that Volterra's publications were destined to be as important for population biology as those of Willard Gibbs for physical chemistry.[4] With mathematical laws to guide their experiments, ecologists could throw off the insecurity they had long felt and begin instead to feel a welcome sense of certainty as far as theoretical possibilities were concerned. All that remained was to map out a program of experiment, observation, and quantification. Ecology was on the way to being an exact science.

Upon his return to America in 1927, Chapman wrote up the results of his flour beetle experiments, wherein he developed the analogy between the growth of the populations and Ohm's law.[5] His article, published in 1928, cited the works of Volterra, as well as Lotka, Thompson, and Pérès, but it is not certain how much he had actually read, for when he began writing to Volterra later in 1928, he requested a copy of Volterra's original article, explaining that he had been unable to obtain it. The article had already been translated into English and published in another journal. Chapman suggested that the paper might be reprinted in a textbook he was writing on animal ecology. Volterra agreed, and the book appeared in 1931 with Volterra's article in an appendix.[6] His promotion of mathematical biology also accounted for the appearance of references to Lotka, Volterra, and Thompson in the foreign literature, for between 1927 and 1931 (when he moved to Hawaii) he invited a number of foreign entomologists to visit his laboratory in Minnesota, where he made sure to introduce them to the latest theories.[7] It was clear, however, that Chapman found the mathematics far too abstruse for his own use, for there was scant discussion of mathematical biology within the body of his book. The highly mathematical appendix stood in stark contrast to Chapman's own research program, which though quantitative was mathematically unsophisticated.

We might smile today at Chapman's naive enthusiasm, but for an older generation of ecologists, his words suggested a line of research which might answer questions that had nagged for decades. When he sketched his ideas before the International Congress of Entomology in 1928, it was Stephen Forbes, the grand old man of ecology, who rose to congratulate him for a "remarkable paper."[8] Forbes reminded the audience that ecologists had often not known when to pause from the gathering of facts to draw a conclusion from the fragments they studied, confident that it would have general applicability. He welcomed Chapman's visions for the future of ecology and declared himself ready to apply "with grateful alacrity, the product which our younger colleagues are contributing to the common stock."

Whereas for economic entomologists the conclusion that population oscillations might be generated by the interactions of the populations came as no surprise, for other ecologists this was a new result. In England, Charles Elton was studying small mammal cycles and trying to correlate them with environmental periodicities, when Volterra's article of 1926 appeared in *Nature*. Julian Huxley, his former tutor at Oxford, brought the article to him. Elton realized its importance. The generation of oscillations through internal causes was new and unexpected. Moreover, he realized that mathematical studies might be useful as a way of generating hypotheses for further field study. But the mathematics was exceedingly difficult, as others had found, and the mathematicians were not always of much help in explaining themselves. Reviewing the first part of Lotka's *Théorie analytique*, in which the mathematical details were considerably toned down in comparison to his other writings, Elton remarked: "Like most mathematicians, he takes the hopeful biologist to the edge of a pond, points out that a good swim will help his work, and then pushes him in and leaves him to drown."[9] But he recognized the potential value of this work, and although he did not adopt mathematical methods himself, he added a brief overview of current theoretical studies to the second edition of his *Animal Ecology* in 1935.[10]

As Bodenheimer observed, the feeling that theory had gone far beyond observation and experiment was combined with an ingrained "mathematicophobia" that was hard to overcome. Grandiose claims fostered among biologists a deep suspicion of theoreticians, the scope of whose assertions seemed matched only by their biological naiveté. Despite Chapman's hopes for a new, exact ecology, his colleagues harbored a mistrust of theoreticians. Would theoretical biologists, they asked, like theoretical physicists, tend to be "rather snooty" and would they "look down upon those of us who expose ourselves to the embarrassment of having to harmonize the data from our observations or experimentation with our hypotheses"?[11] As Eric Ponder remarked, following a talk on the mathematics of growth delivered by Edwin Wilson at the Cold Spring Harbor symposium in 1934:

> I am far from being opposed to biomathematics, but I feel that it is futile to conjure up in the imagination a system of differential equations for the purpose of accounting for facts which are not only very complex, but largely unknown, and the fact that the resulting expressions are not at variance with the observed data really says little for them, unless they are used for descriptive and approximate purposes only. It is said that if one asks the right question of Nature, she will always give you an answer, but if your question is not sufficiently specific, you can scarcely expect her to waste her time on you. . . . What we require at the present time is more measurement and less theory, or, if you like it better, more experimental analysis of phenomena, and less integration.[12]

The counterattack quickly came from Charles Winsor, who had watched the logistic curve dispute unfolding from his inner seat in Pearl's laboratory, no doubt dismayed that, for all the ink spilled, there should have emerged so little enlightenment. He had left Pearl in 1932 to complete a Ph.D. at Harvard in biology. At the same symposium in 1934, he entered a plea for greater use of mathematical reasoning, not as a descriptive adjunct to research, but as a way to begin and to guide more work: "I feel very strongly that it is exceedingly important that the theoretical attack be pushed ahead rather than allowed to wait for the accumulation of observations."[13] He chose to discuss Lotka's and Volterra's work on mixed population growth in order to show that the complexity of nature was not a reason to shun mathematics, rather it was a reason to use mathematics as a way to understand just how complex natural events were: "Such an effort will bring a realization of the number of factors, and of the ignorance we are in as to many of them, such as should completely satisfy the anti-mathematician."

Winsor had little patience with the criticisms that mathematical statements were not "true." Such arguments, as he was well aware, had for the past decade served more to obscure than to illuminate the usefulness of that earlier equation of growth, the logistic curve. The value of the logistic equation, and of the equations describing mixed populations, was that they gave one an entry into the further investigation of nature. They were, in Lotka's words, "animated question marks." If what they predicted was not found in nature, a great deal could still be learned by asking why not. It was in this sense of a working hypothesis that Winsor hoped to demonstrate the proper application of mathematics to population biology.

Despite his feeling that theory should not be held back for want of experiment, however, it was clear that a useful discussion could be conducted only with the aid of facts obtained from both field and laboratory. Winsor had felt his own exposition constrained by the lack of supporting data, especially of an experimental nature, with which to illustrate the mathematical arguments. Much experimental work had been done on populations in the 1920s, two centers of activity being the laboratories of Royal Chapman in Minnesota and Raymond Pearl in Baltimore, but these experiments were either inconclusive or were not directly addressed to the theoretical models. When Chapman, for example, found rhythmic fluctuations in beetle populations that had become infested with a parasite, he thought he had demonstrated Volterra's law of periodic fluctuations.[14] But as Elton remarked, the demonstration of host-parasite fluctuations was by this time nothing new. What was of greater interest was the possibility that these experiments would lead to quantitative measures of the factors involved.[15] Pearl, on the other hand, was not interested in ecological problems. It would be up to the next generation of biologists to work out the ecological implications of his methods of comparative demography.

Experimental research was especially needed to test the validity of the assumptions behind the theories. One of the appealing features of Volterra's discussion was that he had stated his deductions in the form of laws which seemed to be testable by experiment, but the research cited to support these laws did not stand under scrutiny. Egon Pearson, Karl Pearson's son, criticized D'Ancona's interpretation of the Italian fishery statistics, suggesting that perhaps other factors which D'Ancona had not considered, such as changes in the methods of fishing or even migration of the fish, might account for the observations during the war years.[16] Bodenheimer questioned D'Ancona's analysis on other grounds as well, suggesting that environmental factors, such as temperature and salinity, were important influences on the populations.[17] D'Ancona defended his views after reexamining the statistics, but acknowledged that it was impossible to demonstrate a close cause-and-effect relationship between the interruption of fishing during the war and the relative changes in the fish species.[18] However, he still felt the coincidence of the two events strongly supported Volterra's conclusions.

Although Volterra and D'Ancona were able to cite numerous studies of population fluctuations, especially in fisheries biology, in their 1935 monograph, *Les associations biologiques au point de vue mathématique*, such findings hardly constituted proof of Volterra's "laws." Pearson's and Bodenheimer's objections brought out the problems of trying to refer mathematical conclusions to the complex events of the field. At any time, there would be several alternative causal agents to choose from. This variety of possible explanations would make any attempt at proof of a mathematical law inconclusive without a massive amount of data collection. One such attempt at proof by the Russian biologist S. A. Severtzov, who analyzed statistical data from the imperial hunt records, was only partially successful in corroborating Volterra's findings. In particular, quadruped predators did not appear to reduce the number of herbivores in the cyclical manner predicted by Volterra.[19]

It was time to turn to the laboratory and to the younger generation of ecologists, who found in these mathematical studies new ideas for research topics. Chapman's students John Stanley, Frederick Holdaway, and Nellie Payne followed up both mathematical and experimental lines of research on single-species populations and host-parasite interactions.[20] In England, George Salt, having begun under W. R. Thompson's tutelage, continued doing experimental work on parasitism;[21] while in France, Philippe L'Héritier and Georges Teissier, following Pearl's lead, studied fruit-fly populations, as did Bodenheimer in Jerusalem.[22] Also influenced by Pearl was Georgii Frantsevich Gause, a young Russian ecologist whose own teacher, V. V. Alpatov, had spent several months in Pearl's institute in the 1920s. Gause made Volterra's models the subject of his doctoral work in Moscow. I shall defer a more detailed discussion of these interesting experiments to the next chapter.

At Oxford, Charles Elton had just started his Bureau of Animal Population in 1932. This was to serve as a research and information center for the study of population fluctuations in the wild. As research projects got underway, many of them were published in the *Journal of Animal Ecology*, which Elton had persuaded the British Ecological Society to launch in 1932. This was the first journal devoted exclusively to animal ecology, providing space not only for field and laboratory studies, but also for the occasional theoretical paper. It was to Elton's journal that Nicholson sent his first speculative paper of 1933.

The early years at the Bureau of Animal Population were ones of struggle. Money was scarce and came mostly from independent trust funds and private sources. The New York Zoological Society provided an initial grant, but this ended in 1934. It was not until 1936 that the bureau could settle down to steady work, its minimum costs now guaranteed by the university, Corpus Christi College, and the Agricultural Research Council. But with such minimal assurances, existence was still hand-to-mouth, the staff having to scrounge grants from any source they could find.

Most of the bureau's work was nonmathematical, but in 1935, after Patrick H. Leslie (whom everybody called "George") joined as an assistant in biomathematics, a great deal of valuable mathematical help was available. Leslie was not only a good mathematician, he also understood the practical problems of collecting data and designing experiments. He was a mathematician one could talk to; whenever anyone had a problem, he would "take it to George." By working closely with ecologists, Leslie was well aware that an understanding of populations in the wild would require better knowledge of the life tables for these animals. Lotka had provided the starting point with his analysis of stable populations and of the intrinsic rate of increase.[23] Leslie turned this approach to an experimental setting and constructed the first life tables for rodent populations reared in the laboratory.[24] These were, as he said, mere caricatures of the field populations, but their behavior still held some clues about why wild populations fluctuated. He suggested that sensitivity of the breeding cycle to environmental changes might explain the violent fluctuations found in nature.

This experimental start led to further mathematical analysis in the 1940s, much of it an extension of Lotka's work using the methods of matrix algebra.[25] Lotka had finally found his audience in Leslie and the other members of the bureau. In 1948 Elton wrote to him that his ideas were constantly being quoted in discussions of research problems and in the training of advanced students. He requested a photograph to be framed and hung in the library to remind visitors and staff of the importance of Lotka's work for their own research.[26]

These discussions did not quickly yield fruit in the ecological literature, not only because of the difficulty of the mathematics, but because of the enormous practical problems involved in census work on animal populations. Animal demography was still in a primitive state. Elton recognized the important contributions of Lotka and Pearl in his 1942 book, *Voles, Mice and Lemmings*; but, he added, "Since the dazzling intellectual empire conquered by Lotka has mostly lain unexplored by rodent ecologists, and Pearl's ideas were also considered by them as speculative, we may leave them with this very inadequate reference."[27]

Leslie speculated that the constraints of applied research had been responsible for the relative poverty of experimental work addressed directly to these theories. Reviewing D'Ancona's book *La Lotta per l'Esistenza* (1942), he noted with regret that the chapters on experimental and field verification of mathematical theory, representing the research to

1939, claimed only 33 out of 341 pages of text. He hoped that, away from the demands of applied research, ecologists in future might have their imaginations freed to design more creative experiments: "It is an amusing, and a far from unprofitable speculation to consider the way in which theoretical physics might have developed if only an occasional experiment had been done."[28]

The important question was, to what extent should ecology free itself from the applied research which had given it its initial impetus and its continuing support? To what extent should it develop as an academic discipline, seeking to uncover the principles of community structure through careful field research? Who would determine how ecology developed as a discipline: the ecologists teaching in the universities? The theoreticians whose ideas still rested largely unexamined? Or the applied scientists who were concerned to solve particular problems which would give them control over one part of nature?

Mathematical ecology provided not just a source of new ideas, it could also be a source of threat for those whose livelihood depended on an accurate assessment of practical problems. W. R. Thompson was one of those who felt the risk of letting go of biological reality too quickly. Surprisingly, by the end of the 1930s he had become one of the most outspoken opponents of mathematical thinking in biology, opposed to the very trend he had helped to establish and which he now felt had run dangerously out of control.

Mathematical Figments, Biological Facts

Thompson had left France in 1928 to become the superintendent of the laboratory at Farnham Royal, outside of London, of the Imperial Institute of Entomology. The laboratory had been built the year before with a grant from the Empire Marketing Board, a body set up in 1926 to promote the efficient marketing of British Empire products, including agricultural products. The Farnham House Laboratory was to be responsible for coordinating and conducting all the biological control work for the entire empire.

Thompson soon built the laboratory into the largest biological control operation in the empire, although it was modest in comparison with its American counterpart. It had a staff of about five entomologists, a couple of plant ecologists, and a few field workers. As requests from the corners of the empire came in, parasites were bred and shipped off for field testing in those countries. Only the Australians expressed doubts about the wisdom of concentrating all the basic research in British hands. The Australian view was that if they relied too heavily on Britain, then Australian scientists would never get the training needed to raise the scientific standards at home. Economically speaking, British control of research might be the most efficient form of organization, but Australians perceived that it would be suicidal for Australia to allow itself to be dominated by Britain for very long.[29]

Australia was just starting to boost its level of research through the Council for Scientific and Industrial Research, established in 1926. The head of its Division of Entomology was R. J. Tillyard, who joined the C.S.I.R. in 1928 and lost no time applying to the Empire Marketing Board for a separate grant for entomological research in Australia. The Australians also asked that a few of their workers be installed at the Farnham House Laboratory to gain some experience in research. Thompson was not anxious to have these outsiders in his laboratory. He felt that the presence of temporary workers from the Dominions would prevent his building a vigorous research institution.

Despite these mild rumblings, the future at the "Parasite Zoo," as it was called, looked rosy. Biological control was in fashion, so much so that Thompson observed it was "tending to degenerate into a kind of superstition or fad."[30] Previous experiments on biological control had not always been conducted very carefully. Conclusions too often seemed to be based on merely circumstantial evidence and for that reason were viewed with some suspicion by the rest of the ecological community. Thompson hoped to set this part of entomology on a firm scientific basis with a solid program of basic research. Young entomologists who joined the staff in these early years found the atmosphere unusually exciting for an institution outside of a major university, with Thompson at "the peak of his powers as a man of wide and profound knowledge, great industry and much wisdom."[31]

The rosy prospects for expansion quickly faded with the onset of the Depression. In 1931 the British government allocated to the Empire Marketing Board only enough to maintain and to consolidate existing activities. As these economizing measures were passed on to Farnham House, all plans for basic research had to be recast. Projects were limited almost entirely to those financed by the Dominion and Colonial governments, especially Canada and New Zealand. By 1937 it was clear to Thompson that no official support would be forthcoming. In order to attract scientists to undertake the research he had hoped to complete, he was forced to appeal to individual outside workers to come and use the facilities and materials of his laboratory for their research projects.[32]

While Thompson struggled hard to get enough money to keep his laboratory going, events unfolding in related areas of biology posed new threats to his endeavors. Theoretical mathematical arguments were pouring forth, not only in ecology but in population genetics as well: too often they seemed to have little regard for biological reality. Thompson grew alarmed. He began to wonder whether mathematics was not detracting from the importance of long and laborious experimental research in fields such as entomology. Then he recanted. Before the decade was out, he had retracted many of the arguments in support of mathematics that he had made nearly twenty years earlier.

Thompson's fear that mathematical argument could obscure and override biological fact was first awakened by what appeared to be happening in evolutionary biology under the impetus of J. B. S. Haldane and R. A. Fisher. Thompson saw their works as part of a general trend in science away from a direct concern with visible and tangible objects; away, that is, from common sense and ordinary logic. Haldane and Fisher seemed to suggest that the imaginative faculty of the mathematician was as impor-

WILLIAM ROBIN THOMPSON, 1887–1972
Photograph courtesy of Biosystematics Research Institute, Agriculture Canada

tant, if not more important, than the biologist's direct knowledge of organisms. Disturbed, Thompson presented his own views of the relation between science, mathematics, and philosophy in a book, *Science and Common Sense: An Aristotelian Excursion*, published in 1937.[33] As its subtitle indicates, the whole argument was cast in an Aristotelian mold.

Thompson's interests in philosophy went back many years, at least to the early 1920s. A deeply religious Catholic, he had long been a devoted student of Aristotle and Aquinas. In 1924 he had obtained a second doctorate, in philosophy, from the Dominican College of St. Maximin in France. He was well versed in Thomist philosophy, and for many years he carried out a philosophical correspondence with Jacques Maritain, who wrote the preface to his book. These beliefs were woven densely into the fabric of his thought. As a former student recalled, "It was difficult to talk to Thompson about any subject not purely technical without the 'Angelic Doctor' creeping into the conversation somewhere."[34] Within this framework, Thompson's latest position was based on the argument that mathematics as a deductive science had to be carefully distinguished from the essentially inductive natural sciences. He argued that deductive reasoning, as used in mathematics, referred only to ideal entities, whose properties were determined by the original definitions or axioms. But biological science was concerned with cause-and-effect relationships in the real world; such relationships could not be deduced, but could only be discovered by laborious observation, experiment, and induction.

The crucial word was "laborious." Thompson was afraid that too great a reliance on mathematics would draw attention away from the need for detailed, long-term field studies. The mathematicians seemed to suggest that the kind of work he was engaged in, the work which he had had to fight to keep alive for the past few years, was not the best route to new knowledge. In *The Genetical Theory of Natural Selection*, Fisher had urged biologists not to scorn the ability of mathematicians to think in abstractions, for through abstract reasoning they would come eventually to a better understanding of biological reality. The mathematical method, according to Fisher, consisted of abstracting the essential elements of a problem, which was considered to be "one of a system of possibilities infinitely wider than the actual." Generalized reasoning allowed one to grasp the essential relations of those possibilities. If these relations could be subsumed within a formula, one could, armed with the formula, return to the consideration of particular cases. As Fisher wrote, by way of example, "No practical biologist interested in sexual reproduction would be led to work out the detailed consequences experienced by organisms having three or more sexes; yet what else should he do if he wishes to understand why the sexes are, in fact, always two?"[35] To which Thompson, not amused, replied: "If the imaginary animal became a real animal, he might turn out to be quite different.... We cannot show by considering an imaginary animal, why real animals are as they are. The only possible way of discovering this is to consider the real animals themselves."[36]

There were other reasons for Thompson's displeasure. Fisher and Haldane had both used mathematics to make an argument for natural selection as the principal mechanism of evolution. But Thompson did not believe in natural selection; indeed, he felt that the very idea that the diversity of life could be explained by natural forces and interactions was only a hypothesis, still in need of proof. The "energetic young geneticists and statisticians" had acted much too rashly in assuming natural selection's importance before the experimental evidence had been completely gathered. The physiological mechanisms underlying adaptation still had to be demonstrated. The Darwinian theory was, in his view, not scientific but philosophical; it was a plausible fiction, a simple deduction, "curiously allied" with mathematics because of its deductive structure, yet with insufficient experimental basis to warrant its acceptance.

Thompson's point about experiment and observation was not unusual for the time, for he was part of the generation that had rejected the speculative morphological reasoning prevalent in the late nineteenth century, and had sought to replace it with an experimental method which would lead to positive knowledge. But his arguments also concealed a philosophical, religiously motivated antipathy toward any theory of evolution (although he allowed that the Lamarckian hypothesis was at least intelligible). He could not accept the view that nature was infinitely plastic. Clearly, the geological record indicated a progression of organic types, but this was all that could be asserted. Thompson believed that all species possessed an essence, or form, in the strict Aristotelian sense, which could not be changed by material means. He had a strong aversion to the hypothesis that all adapted types had been produced by the random actions of mutation and natural selection. Such random processes might explain microevolution, but they could not explain macroevolution. Thompson could not suggest an alternative hypothesis for the ones he had rejected; he simply regarded evolution as an unsolved problem.[37]

His efforts to discredit these theories of evolution on logical grounds led him to a position of extreme skepticism, to a rejection of any theory or concept for which positive factual evidence was lacking. If a hypothesis could not be tested rigorously, experimentally, then there was no reason to accept it or to use it in scientific discourse. His scientific writings were strikingly devoid of metaphor, analogy, or theory. It should not be concluded, however, that Thompson was a positivist, although his skeptical point of view was certainly influenced by positivist philosophy. But Thompson rejected positivism as too narrow because it removed all philosophy from science. He believed that philosophical concepts, such as evolution, did have a place in science; without them, science became dehumanized, "a mere cinematographic record of events."[38] His argument in this respect contained the same basic inconsistency that characterizes modern "scientific creationism." On the one hand, Thompson accepted the importance of philosophy, of an element of belief, in science; on the other hand, he resorted to a rigid insistence on the facts, on experimental testing, on positive knowledge, when confronting theories which went against his religious grain. This point of view made Thompson a keen critic, but one who was not able to construct anything in place of the theories he had criticized.

Population ecology soon received the benefit of his critical eye as well. What alarmed him especially was not the continued growth of mathematical ecology per se, but rather his feeling that people had stopped discriminating between mathematical figments and biological facts. As long as Volterra had referred to his research as a study in pure mathematics, certain excesses of metaphorical language could be excused. But lately Volterra's writings had seemed to become "more positive and dogmatic" in tone.[39] When in 1937 Volterra claimed that "the laws which proceed from the general equations have been verified by several biologists," Thompson felt that things had gone too far. Mathematical ecology was no longer just a stimulus to the imagination, a way to help biologists envisage a problem. People were beginning to believe that it held the truth.

Nicholson and Bailey were particularly guilty of a willingness to believe their own arguments. There was much cause for disagreement with the way Nicholson had used analogies to illustrate competition without ever demonstrating that competition really occurred, or that there was a "balance of nature." Moreover, their arguments contained too many simplifying assumptions that Thompson knew were not supported by biological observation. His criticisms echoed those of the entomologists who had responded to his own first mathematical efforts: as long as the equations included parameters that could not be calculated, or assumed that some factors were constant when in reality they were variable, then there was no chance of applying the mathematical results to field research.[40]

But the argument did not rest long on a quarrel over facts. Thompson accused Nicholson of more general errors in reasoning and judgment in the use of mathematics. In combating Nicholson, Thompson was now forced to reverse his own youthful views about mathematics. He admitted that he had failed earlier to appreciate the significance and the limitations of mathematical methods. He warned that economic entomologists were falling into the same trap into which he had fallen: the mistaken belief that a mathematical relationship implied a real cause-and-effect relationship between natural populations. He pointed out that an equation predicting an oscillation of natural populations was a mathematical deduction describing only an ideal case. It implied nothing about the actual causes underlying the oscillations of the real populations. To assume that the equations implied true causality between host and parasite populations was to confuse the ideal mathematical entity with the real object.

Thompson would still allow some mathematical analysis in entomology in order to describe biological relations or even to suggest further experimental work. He was most concerned, though, that entomologists should understand that mathematical "laws" merely expressed natural processes, they did not govern them. The distinction he made between a law as description versus a law as explanation was similar to that made by Karl Pearson in *The Grammar of Science*. But where Pearson regarded such laws as having predictive value in physics, Thompson denied that they could ever do so in biology. This was not just because mathematical expressions were too simple in comparison with nature. Mathematics, no matter how intricate, could never help biologists to understand causal relations, because natural events were made up of unique causes which were necessarily unpredictable:

> The tremendous multiplicity of factors acting on the real world has not merely the complexity of an elaborate mathematical equation, which is theoretically but not practically manageable; but implies a genuine unpredictability because the actual combination of factors has never been observed to operate and until it has, we cannot really be *sure* what its effect will be. Much less can we *see* this effect in its causes.[41]

Thompson realized it was not correct to assert that such events were absolutely unpredictable. If the causes were independent, if they could be known separately, and if the point of their intersection could be determined, then the event could be predicted. "Nevertheless," he concluded, "we cannot strictly speaking build up a science of the fortuitous," any more than it would be possible to make a science out of history.[42] Entomology was still a science, but it was a science not of populations but of individuals, Thompson seemed to say. It had to be based on observation and experiment, on knowledge of how individual organisms behaved. Apart from social species, the population as a unit had no existence; therefore it could possess no self-regulating properties, no mechanisms to adjust to a hypothetical balance of nature. Natural control was to be understood, not in terms of mystical regulating factors of the kind Nicholson proposed, but by understanding the physiology of the organisms themselves.[43] Thompson objected to the way mathematics had seemed to imply that the population behaved as a unit governed by mathematical laws.

Thompson's reaction was certainly justified in view of the vague metaphorical language that Nicholson had used to build his argument. It was also justified in the light of the tendency of some ecologists, such as

Chapman, to try to turn ecology into a physical science with its own laws of nature. Thompson's point of view expressed the biologist's concern with individuality and with history. The same argument would be made much later in 1959 by Ernst Mayr in response to the rise of mathematical population genetics: "The more I study evolution the more I am impressed by the uniqueness, by the unpredictability, and by the unrepeatability of evolutionary events." Mayr continued, "Let me end this discussion with the provocative question: 'Is it not perhaps a basic error of methodology to apply such a generalizing technique as mathematics to a field of unique events, as is organic evolution?' "[44]

To point out that Thompson had justification for his arguments does not quite explain the timing of his change of mind, or its vehemence. For his new arguments not only reversed what he himself had believed, they became unusually exaggerated as time went on. By the late forties he had adopted all the uncompromising rigidity of a man who felt afraid. And although he never openly said so, he had cause to feel a particular threat in this instance, for Nicholson was now a direct rival.

In 1930 Nicholson had joined the Division of Entomology of the Council for Scientific and Industrial Research, where he shortly had to take full administrative control over from Tillyard, who had become ill. In 1936 he was made chief of the division. As a result of Tillyard's energy and foresight, the division was receiving support from the Empire Marketing Board. Thompson and Nicholson, therefore, were having to share the same source of funds during this period of economic crisis. Moreover, Thompson was dismayed to see that Nicholson's arguments were converting his formerly skeptical colleagues. Harry Smith, who had not been won over by Thompson's own arguments in the 1920s, was enough taken by Nicholson's later on that he began to wax enthusiastic about the potential of mathematics in biology. In 1939 he was arguing, like Chapman, for a more vigorous program of experimental testing: "In that admirable work by Nicholson and Bailey, 'The Balance of Animal Populations,' will be found enough population problems to keep several laboratories busy for the next twenty years."[45] Smith and his graduate student, Paul DeBach, performed some of the first experiments, published in 1941, aimed at verifying Nicholson's predictions.

Thompson meanwhile was having to appeal for researchers to carry on the important work in biological control for which he could not get funds. The war made his position even more precarious. All field work in Europe stopped, the laboratory at Farnham Royal was closed, and Thompson was transferred to Canada to run the Imperial Parasite Service. Soon funding for this office was also cut back, and again his plans for expansion were frustrated. It was only with the creation of the Commonwealth Institute of Biological Control in 1948 that he achieved any guarantee of stability. It is no wonder that, seeing his own position eroded gradually, he should have been aroused at the prospect of entomology becoming the willing servant of a theory which he felt was based on errors both of fact and of reasoning.

Thompson wanted to warn that theory ought not to replace the study of nature, the laborious field research necessary to any sound planning. But this was surely a warning that no biologist, no matter how theoretically minded, would dispute. Nicholson said nothing at first, then after years of silence spoke out in support of his methods. Bemoaning the attitudes of biologists such as Thompson, he pointed out in 1954 that science consisted of the continual interplay between thought (or theory), observation, and experiment: "The common and reiterated insistence upon the paramount importance of observation and experiment, and the deprecation of 'theorizing' (which seems to be the fashionable word for any deliberate and sustained thought) indicates a gross misunderstanding of scientific method."[46] Nicholson realized that any deductive argument depended on the truth of the postulates underlying it, but he felt that if field observations did not support mathematical conclusions, this merely indicated that the postulates were incorrect, not that the deductive method was itself wrong. The solution would be to modify the postulates in accordance with known fact, and to begin again with new hypotheses.

His own methods did move toward such a balance of theory, experiment, and observation, but only after a delay of more than a decade from his first article of 1933. The delay was caused by a combination of lack of time and lack of inspiration. His administrative duties prevented him from engaging in much research until after the war. But he had also assumed it would be impossible to test his theories in the laboratory, because the postulates had required large populations and a vast amount of searching space for the parasites.[47] DeBach and Smith's article of 1941 showed him otherwise.[48] They had set up experiments using housefly puparia and their hosts, placed in large tins filled with barley. From their example, Nicholson started his own series of experiments using sheep blowfly cultures, studying how oscillations were created in the populations, and later observing the results of natural selection in the populations.

By the 1950s, when Nicholson finally rose to make his defense of theory, Thompson had already eased his attack on mathematics, though he remained cautious (and on natural selection and the nature of species did not waver). In his later years, Thompson looked back with pride to his accomplishments in pioneering the use of mathematics in ecology. But others were not going to let Nicholson win an easy victory. The whole problem of the utility of mathematical models in relation to the factors controlling populations and in relation to the importance of competition was hotly debated among ecologists throughout the decade and well into the 1960s.

In chapter seven, I turn to a slightly different issue which also pertains to Nicholson's work on competition and adaptation: the relation between ecology and evolution. But first I must consider another question raised by Thompson's views: the position of mathematical population genetics, itself flourishing in this period, in relation to the ecological studies discussed so far. This too is relevant to the question of evolution and its place in ecology.

Population Ecology and Population Genetics

From Thompson's medieval perspective, population genetics appeared a misguided venture into mathematics to be shunned because it carried the baggage of the unproven idea of evolution by natural selection. For most others engaged in population ecology, these recent genetical studies were simply not very relevant because they did not take into account ecological relationships. Moreover, much of mathematical population ecology began by assuming that populations were not evolving; an assumption made necessary by the complexity of population interactions.

Lotka was certainly not interested in the evolution of species, though he was well aware of current research in genetics. His entire discussion of the evolution of species consisted of a lengthy quotation from Haldane's early work. He then dismissed the problem in order to concentrate on the "evolution" of the whole system. Similarly, although there were points of contact between his analysis of the intrinsic rate of increase, r, and Fisher's actuarial approach to genetics, there was no relation between Fisher's genetical theory of natural selection and Lotka's program of physical biology. Later on, others influenced by Leslie's work would draw upon both Lotka and Fisher to make the statistic r a measure of Darwinian fitness. But the use of demographic techniques in an evolutionary context would not be seriously entertained by ecologists until the 1950s.

Volterra was more directly interested in evolution, that is, in the analysis of Darwin's theory of the struggle for existence. But his method began by supposing that evolution did not occur, that the population was a homogeneous aggregate, unvarying through time. As a problem in the mathematical modeling of short-term fluctuations, a sufficient justification of the effort was that the models did appear to approximate those changes seen in nature. Longer-term evolutionary changes had to be excluded at this preliminary stage of the analysis in order to simplify a method that had already become highly complex. These studies were not unknown to the geneticists, but they were considered to bear upon different problems than those the geneticists were trying to understand. Haldane referred to the "beautiful work" of Volterra and Lotka, only to dismiss it as not germane to his interests.[49]

The same assumption of homogeneity and invariance could be seen in

Bailey's argument, which adopted much the same point of view as Volterra's. Although Nicholson connected his theory to natural selection and speciation and was aware that genetic composition was undergoing change, the necessity of simplifying the mathematical argument precluded any detailed consideration of genetic variance. Indeed, he did not know enough about his populations to integrate genetics into his analysis.

In all of these cases, attention was focused on short-term changes and the mathematical representation of those changes. The span was ecological rather than evolutionary time. Moreover, the unit of study, the population, was not distinguished by any biological, genetic, or geographical characteristics which might make it meaningful as an evolutionary unit. In mathematical ecology, the population was as artificial an entity as a population of molecules in an enclosed container, the product of the physicist's rather than the biologist's cast of mind. The study of populations from this viewpoint did not include what Ernst Mayr has identified as the crucial requirement for an evolutionary perspective, "population thinking," that is, the perception of a population as composed of genetically unique individuals.[50] The only exception among the theoretical group of the 1930s was V. A. Kostitzin, who combined Lotka's and Volterra's methods with an evolutionary and genetical awareness in a series of articles published in the late 1930s.[51] Unfortunately, these did not receive wide notice, and Kostitzin's biological research in this area largely ended during the Second World War.[52]

Even among field biologists who studied populations, genetics was slow to find a place. In the 1930s this was hardly unusual, for genetics had not permeated to any great extent the other biological disciplines. E. B. Ford, for instance, recalled how exceptional were his own interests in ecological genetics at Oxford in the late 1920s.[53] Early signs of a synthesis of the genetical and naturalist viewpoints were apparent in the 1920s, but the full evolutionary synthesis did not gather its momentum until the late 1930s.

Although it was generally believed that "evolution was long and ecology short," some ecologists did sense that ecology and genetics had something to offer each other.[54] Charles Elton kept arguing into the 1930s, much as he had done a decade earlier, that the study of population fluctuations, of how they influenced the spread of mutations and how they affected adaptation, was one way to integrate these separate disciplines. But the synthesis of ecology and genetics was still a long way off. Even when it was recognized as an arbitrary distinction, it was maintained, as Thomas Park said, in part out of convenience in analysis and in part reflecting the competence of the worker.[55] Only in the 1960s did the two fields of population ecology and population genetics begin to merge under the general heading of population biology. The joining of these two perspectives is still far from complete.

But though genetics and ecology would remain aloof for some decades yet, owing to the tremendous complexity of an integrated approach, it was nevertheless true that in the 1930s ecologists were becoming more interested in problems of evolution, especially of natural selection's role. Ecology was not considered to be an evolutionary science; its main goal was still to understand communities. But as more was learned about the patterns of distribution and abundance, it became clear that this information could be turned to answer evolutionary questions. The wide-ranging viewpoints of the biologists who were emerging as the architects of the modern synthesis in evolution between the wars began to exert an effect on ecology as well as the other biological disciplines.

One of the results of these developments was a revival of interest in competition in the 1940s. I do not wish to imply that competition was in any way a new issue; ecologists, especially plant ecologists, had always been interested in competition. But within population ecology it rapidly took on a new prominence among animal ecologists, overshadowing the companion problem of predator-prey relations for a time. In the study of competition the findings of mathematical ecology, laboratory ecology, and field ecology were merged, starting with the simple and elegant experiments of G. F. Gause in Moscow. The underlying problem, however, was an old one for ecologists: the problem of how similar species could live together, in theory and in practice, and what this coexistence implied for the structure of ecological communities.

7

The Niche, the Community, and Evolution

Georgii Frantsevich Gause has become famous in the history of ecology for his enunciation of a simple maxim, that two species living together cannot occupy the same ecological niche. The "principle of competitive exclusion," as it is now called, was by no means unheard of when Gause stated it in 1934. In fact it had been expressed often in the naturalist literature and was felt to be obvious and therefore unremarkable. Gause himself thought so the first time he stated it. Yet within a few years it became a cause célèbre among ecological ideas, a rallying point for the ecological study of competition. Its importance owed nothing to the novelty of the idea, but to the fact that it nicely complemented ecologists' renewed interests in competition and evolution. Though Gause eventually concluded that his principle was significant as a way to organize ideas about community structure, the establishment of his principle as a leading idea of ecology sprang not so much from his own use of it as from the impetus it received from George Evelyn Hutchinson and David Lambert Lack. Hutchinson used the principle to explore the meaning of the niche; Lack to explore some related problems in evolutionary biology. Gause himself was led to his principle through his experimental studies of Volterra's equations and came to these in turn through the influence of his teacher, Vladimir V. Alpatov. Alpatov had studied with Raymond Pearl in the late 1920s and had enthusiastically absorbed Pearl's methods and ideas. Gause brought Pearl's methods and Volterra's ideas together to answer an ecologist's questions.

Moscow Ecology and Baltimore Biometry

Vladimir Alpatov had been interested in problems of geographical distribution and variation in invertebrates. To prepare himself for research, he began to study biometry and taught a course in the subject at the Zoological Museum of Moscow University. Through one of the professors there, Aleksander Gurvich,[1] he came across a few of Pearl's papers. From his perusal of Pearl's work, he concluded that the best way to complete his education was to visit Pearl in Baltimore:

> I cannot but express to you my intense desire to obtain the means of working on experimental and biometrical themes under your direction. Not only in

> Moskow, but nowhere in Western Europe, can I hope to acquire the knowledge and experience, which working near you at the Institute is sure to give. It was with the greatest difficulty that I could obtain here the publications of your laboratory.[2]

With a fellowship from the International Education Board of the Rockefeller Foundation, Alpatov set out for America in 1927. He spent the summer of 1927 at the laboratory of entomologist E. F. Phillips at Cornell University, where he continued his studies on the honeybee. From Cornell he moved to Pearl's laboratory in the fall and was soon engaged in the ongoing work on *Drosophila* under Pearl's direction. A renewal of the fellowship allowed him to stay in America until August 1929. During this year he pursued the *Drosophila* studies, working on the influence of temperature and starvation on the physical constitution of the flies. He

Georgii Frantsevich Gause, b. 1910
Photograph courtesy of G. F. Gause

returned to Russia much impressed by the way science was carried out in America, where the biologist could devote his existence to research on a scale that was impossible in Russia at the time.[3] But above all it was Pearl himself who had most impressed the young man, as he explained in his report to the Rockefeller Foundation: "My personal contact with Doctor Pearl during the two years in Baltimore ought to be considered as the most important achievement of mine in this country. It seems to me more important than the experimental work done and the problems I succeeded in solving here."[4]

Having intended to delve deeper into biometry when he first made plans to come to America, he returned to Russia eager to follow a more physiological, experimental line of research. He attributed this change in direction to his Baltimore experiences.[5] With the Russian emphasis on applied science, he had to give up the idea of working on fruit flies and to concentrate instead on animals of greater economic importance: silkworms, bees, and moths. Apart from physiological work, some of his projects paralleled those in Pearl's laboratory, such as the influence of density of population on the growth of silkworms and wax mothworms, and the duration of life in butterflies.[6] Later he studied the influence of ultraviolet radiation on silk production and on disease in silkworms.[7] On the whole, however, Pearl's influence was not manifested in the subject matter of Alpatov's research, but in the enthusiasm for experimental biology generated in the beginning biologist. This enthusiasm in turn affected Alpatov's bright young student, G. F. Gause, with notable results.

While Alpatov was in America, Gause was engaged in ecological research on animal abundance in relation to habitat. He spent the summers of 1928 and 1929 gathering data on the distribution of grasshoppers in the Northern Caucasus and published the first year's results in *Ecology*.[8] When Gause heard of Pearl's work from Alpatov, he quickly grasped the relevance of the experimental method to his own studies on the correlation between population and environment. In the field it was only possible to correlate abundance with the whole microclimatic complex; under the simpler conditions of the laboratory, he felt, it would be possible to determine accurately how a specific ecological factor influenced population size. After a preliminary exploration in this direction, using data from other published sources,[9] he expanded his ideas on the basis of his own investigations and sent an article to Pearl, who published it in his *Quarterly Review of Biology*.[10] At the same time, Gause and Alpatov jointly wrote a review article on the logistic curve, which explained the theory and cited various applications of the curve.[11] The article, published in German and Russian, was intended primarily to publicize the curve in these countries.

In the meantime, Alpatov was trying to find a way that would enable

Gause to study with Pearl in America. Pearl had suggested applying for the same International Education Board fellowship that Alpatov had received, but competition from Russian candidates was very heavy and his chances were slim. Despite a strong letter of recommendation from Pearl[12] and a personal visit to Alpatov's laboratory by a representative of the Rockefeller Foundation's Paris office, who was much impressed by Gause's work,[13] their request was turned down in 1932. The only objection given was Gause's age: he was at that time just turning twenty-two.[14]

Thinking that the publication of a book in America might enhance his chances for the fellowship, Gause proposed to Pearl a work based on his latest population research. Pearl responded favorably, but warned Gause about the caginess of American publishers in a time of depression; he could not expect to make any money from the book.[15] Gause went ahead and sent Pearl the manuscripts. After some delay on Pearl's part, they were ready for publication in 1934. He still failed to get the grant, but his lack of fortune was in no small measure compensated by later fame, for his small book, called *The Struggle for Existence*, eventually became a landmark of experimental population ecology and a testimony to his originality.

Gause followed up these preliminary studies with a second book, *Vérifications expérimentales de la théorie mathématique de la lutte pour la vie*, published in 1935 in the same series edited by Georges Teissier to which Lotka, Volterra, and Kostitzin also contributed. These studies formed the basis for his doctoral thesis, presented to the University of Moscow in 1936. As the title of the second book indicates, his experiments were designed to test the predictions of Volterra's mathematical models in associations of two species. Volterra's methods had indicated a way to quantify the Darwinian idea of the struggle for existence in terms of the numbers of individuals in mixed aggregations. By placing two species together in a limited environment under different competitive and predatory arrangements, Gause could see whether the shift in the numbers of individuals conformed to the predictions of the models.

The experimental subjects were yeast and protozoa. Although he had started with fruit flies, using populations that Alpatov had brought back from Pearl's own laboratory, he soon switched to microorganisms, which were easier to handle in the laboratory. There was a precedent for using yeast populations: Oscar W. Richards at Yale University had recently completed a study of yeast growth and its mathematical representation for his Ph.D. thesis.[16] His studies were meant to refute T. B. Robertson's growth hypothesis, still a center of some controversy, but they had a bearing on Pearl's logistic hypothesis as well. Gause used these studies as a basis for his own growth analysis, although his purposes were totally different from Richards's.

Gause was interested in the experimental study of ecology and evolution. A problem of great interest was how the relatively stable structure of the community, or biocoenosis, arose through the dynamic interactions of the populations in the community. This area of ecology had been greatly stimulated in Russia by the holistic systems approach developed by V. I. Vernadsky. He perceived an individual species of the community in terms of its role in the cycling of material through the system. The place of a species in the economy of nature could best be understood by looking at the large picture of biogeochemical cycles throughout the entire world system. Seeking to extend this method of analysis further, other ecologists in Russia, such as V. V. Stanchinskii, developed a more mathematical, energetical approach to ecological systems.[17] Gause himself worked in Vernadsky's laboratory in 1933, and Vernadsky served as one of the official examiners of his thesis. Like their American counterparts, these ecologists were hoping to turn ecology into an exact, quantitative science. Related mathematical methods in the foreign literature were received with much interest. Already Pearl's researches were being cited by the Russians, one of the earliest citations being in Vernadsky's book *The Biosphere*, published in 1926.[18] Volterra's articles were also well known. Russian ecology was a thriving, innovative science, looking toward a future of comprehensive systems analysis and quantitative methods, anticipating in some areas the trends that would soon develop in American ecology as well.

The Population as Computer

To test the predator-prey model, Gause set up populations of protozoans in test-tube environments.[19] The Lotka-Volterra equations had predicted that the two populations would oscillate continually, without either species going extinct. Gause's populations were not so obliging: his predators quickly consumed all the prey and then died off shortly afterward. The Lotka-Volterra equations, with their too simple assumptions about the nature of the encounters between predator and prey, failed to take into consideration the biological properties of the species involved. Instead of the "classical" oscillations of the equations, Gause obtained what he called "relaxation" oscillations, or extinctions; the term "relaxation" referring to an abrupt change analogous to the abrupt changes in wave form of electrical relaxation oscillators. This result led him to revise the original equations so that they predicted "relaxations" or extinctions in conformity with the observations.[20]

But were true oscillations possible under other conditions? Gause tried introducing periodic "immigrations" into his populations and found that indeed oscillations of the two populations did occur. This suggested that any predator-prey oscillations observed in nature might not be the results of the interactions between the two species alone, but might occur because of constant interference from without. The findings of his Russian colleague S. A. Severtzov, that predator and prey populations did not fluctuate in a regular manner in some cases, seemed to support these experimental conclusions.

The predation studies opened some interesting questions for the progress of evolution in nature.[21] Gause found that when predation was at a very low intensity, oscillations of the Lotka-Volterra type did arise. But if the predator evolved a more efficient method of capture, the intensity of predation would increase, and the oscillations could be expected to transform themselves into relaxation oscillations. If, on the other hand, the evolution of the predator were balanced by the evolution of better defenses in the prey, then the system would again favor the appearance of classical oscillations.

These hypotheses led Gause to a general principle: "The increase in the adaptability of the aggressors tends to transform associations into relaxation associations, and the same increase in adaptability of the victims brings them back again to associations of classical fluctuations." The striking use of the words "aggressors" and "victims" hinted at wider applications to human societies, but if Gause had any intentions of elaborating this principle, he did not let on. With this sentence he ended the *Vérifications expérimentales*, and his studies of evolution among predators and their prey were not developed further.

The competition model, on the other hand, received more attention. Volterra had derived a simple model for competition between two species which was actually a modification of the logistic equation. He had not known of Pearl's work at the time, but had derived the equations easily enough from first principles. Gause brought the two together, setting Pearl's logistic theory as the starting point for the study of competition. He waved aside the criticisms that had been aimed at the logistic equation so far: "We must not be afraid of the simplicity of the logistic curve for the populations of unicellular organisms and criticize it from this point of view," he wrote. "At the present stage of our knowledge it is just sufficient for the rational construction of a theory of the struggle for existence, and the secondary accompanying circumstances investigators will discover in their later work."[22] He readily concurred with O. W. Richards that the curve did not reflect the complexity of growth and was therefore only an approximation, but he felt this should not prevent analysis of those cases where the curve did appear to describe the course of a population. For one-celled organisms in test tubes, it seemed perfectly adequate.

The logistic equation provided a way to express the struggle for existence quantitatively in terms of numbers of organisms. Moreover, Gause noticed that it could be linked very neatly to the concepts of biotic potential and environmental resistance that Pearl's rival, R. N. Chapman, had coined. He illustrated the connection by the following word-equation:[23]

$$\left[\begin{array}{l}\text{Rate of}\\ \text{growth or}\\ \text{increase per}\\ \text{unit of time}\end{array}\right] = \left[\begin{array}{l}\text{Potential}\\ \text{increase of}\\ \text{population per}\\ \text{unit of time}\end{array}\right] \times \left[\begin{array}{l}\text{Degree of}\\ \text{realization}\\ \text{of potential}\\ \text{increase}\end{array}\right]. \quad (7.1)$$

$$\frac{dN}{dt} = rN \times \left(\frac{K - N}{K}\right). \quad (7.2)$$

Chapman's biotic potential was represented by the product rN, while the degree of realization of this potential, $(K - N)/K$, depended on the environmental resistance. He considered the actual measure of environmental resistance to be $1 - (K - N)/K$. He did not resort to the Ohm's law analogy, but thought of the closed microcosm as having a certain number of available places which the individuals filled as the population grew. The environmental resistance was a measure of the number of vacant places left to be filled.

Having given Chapman's ideas quantitative meaning by connecting them to the logistic equation, Gause proceeded to analyze how environmental resistance limited the biotic potential of a population. He found in his experiments on yeast that he could measure the environmental resistance by the amount of toxic wastes produced by the growing yeast populations. Because he had been able to calculate this resistance independently of the logistic equation, he believed that the experiments confirmed that the logistic curve did express the mechanism of growth. It could therefore be regarded as a law of growth. His results were not as conclusive as he implied. All he was able to show was that, in a mixed population, one species will inhibit another in proportion to its production of toxic materials. Such experimental confirmations are open to the criticism, recently expressed by G. E. Hutchinson, that they may not actually reveal much more than is already known about the population:

> What we have indeed done is to construct a rather inaccurate analogue computer for giving numerical solutions of our equation, using organisms for its moving parts. When we find that we have confirmed the logistic, what we have mainly confirmed is that the reduction in the rate of population growth is linearly dependent on the relative density of organisms. Actually, the beautiful S-shaped integral curve may be too insensitive a result to tell us how well we have established this conclusion.[24]

But Gause had reasons for maintaining his faith in the logistic equation, for it was also the basis of Volterra's competition model. Competition between two species could be expressed by the following equations, in which the normal logistic growth of each species is modified by a term representing the competitive interaction between them:

$$\frac{dN_1}{dt} = b_1 N_1 \frac{K_1 - (N_1 + \alpha N_2)}{K_1}, \tag{7.3}$$

$$\frac{dN_2}{dt} = b_2 N_2 \frac{K_2 - (N_2 + \beta N_1)}{K_2}, \tag{7.4}$$

where N_1 and N_2 represent the number of Species 1 and Species 2, respectively. The coefficients, α and β, express the struggle for existence between the two competing species: α represents the intensity of the influence of Species 2 on Species 1; β, the intensity of the influence of Species 1 on Species 2. The equations predicted that when two species competed for the same energy source, one would always drive the other to extinction. By setting up mixed populations of protozoans in such a way that they competed for the same food source, Gause found that he could fully confirm this prediction. Competition always resulted in the extinction of one of the species.

The necessity of competitive extinction applied only to highly simplified microcosms, however. Lotka, on the contrary, had thought it unlikely that two species could have the same diets in the real world. Stimulated by Volterra's results, he took up the question briefly in 1932 and reanalyzed the problem after eliminating the condition of identical diets.[25] Lotka's new solution yielded an additional point of equilibrium which he felt might correspond to the case that both species survived: "This is more in keeping with the facts of nature, since it is a matter of the most common knowledge that a great variety of species of organisms sharing certain resources of food do live together in essentially stable equilibrium."

It was characteristic of Lotka's interests in human society that, having modified Volterra's results, he saw the treatment as more applicable to economics than to biology. Recalling the nineteenth-century analysis by Augustin Cournot of economic competition, he singled out a contemporary criticism that Cournot's analysis implied that any competitor with a slight advantage over the others would eventually displace them and become a monopoly.[26] But, Lotka argued, just as competing animals had slightly different diets, so economic competitors might have different access to their sources of supply and to their markets. What would occur would be a scattering of competitors over a given region, each having superiority in its own territory, but relinquishing the advantage to another competitor in a different area. Coexistence was therefore wholly possible through partitioning of territory and resources. As for biological competition, Lotka was less interested. He thought the possibilities of finding concrete examples in nature corresponding to the equations were remote, although the laboratory promised greater success in setting up the requisite conditions. He concluded simply that "it would be interesting to see the experiment actually made."

Gause, having confirmed Volterra's conclusions in the simplest possible environment, next turned to a more complicated interaction of the type suggested by Lotka's remarks. These experiments were reported in his French monograph. He populated his microcosm this time with two competitors having slightly different ecological requirements, and with two food sources, yeast and bacteria. The one species, *Paramecium bursaria*, contained symbiotic algae which allowed it to live comfortably in an oxygen-poor environment. It tended to keep to the bottom of the test tube, where bacterial action had depleted the oxygen, and fed mostly on yeast cells which deposited themselves there. *Paramecium caudatum*, on the other hand, preferred the upper layers and fed mostly on bacteria. Each species was capable of eating both bacteria and yeast, but each kept to a section of the tube where it could profit more easily from one food source than another. Competition therefore enabled the two species to coexist by forcing them into separate parts of the environment, or into separate ecological niches. This was the biological counterpart to Lotka's example of economic competitors.

Gause realized that the predictions of the models were valid only in a qualitative sense, that is, in terms of general trends such as extinction, coexistence, oscillation, and so on.[27] His discussion of how these models might be used was an improvement over Chapman's notion that the experiment could unequivocally "prove" the mathematical "law" which could then stand as the basis for an exact science of ecology. Gause understood that validating these kinds of models could not be a question of such simple, linear thinking as Chapman proposed. Instead, the relationship between the model, experiment, and outcome was reciprocal. The model imposed certain limitations on the conditions of the experiment, and the experimental results in their turn indicated what changes, if any, had to be made in the model to make its predictions conform to experimental reality. There was in this way a continual feedback between the model with its cluster of assumptions and the experimental results. To use Hutchinson's analogy once more, the population is treated as a computer where a mathematical program is translated into biological language. But like the logistic curve, these models were too insensitive to enable the experimenter to conclude that all the assumptions behind the models were fully confirmed whenever the general predictions of the model were fulfilled. Gause realized this and stressed that any quantitative theory of population growth would have to be "confirmed by the data obtained through entirely different methods, by a direct study of the factors limiting growth."[28] He had hoped to provide this independent confirmation in his studies of logistic growth in yeast.

There is always the possibility, when this type of reciprocal relation is involved, that the whole complex is a tautology which, as Hutchinson points out, may not reveal anything that the experimenter does not already know. As we shall see in chapter eight, a large portion of Hutchinson's writings were meant to show that such need not be the case. These experiments can be useful if they are taken to be the first step in the construction of different hypotheses more directly referable to nature. Hutchinson was seeking not only a better understanding of the complexity of nature, but better insight into the larger patterns and sequences of events which would create a more unified understanding of the world. From these specific hypotheses, he would move toward a general theory.

Gause also understood this function of his experiments. For both competition and predation he tried to relate his results to broader issues. The competition experiments were connected to observations about the structure of ecological communities; the predation experiments to observations about the mechanisms of evolution in biological aggregates. The value of his books was not just in the description of the experiments, but in his ability to relate them to larger ecological questions and to a broad range of ecological literature. The approach pioneered by Gause would later be extended by Robert MacArthur in the 1960s. MacArthur himself seemed not to appreciate the full extent of this debt, for it came about indirectly through the influence of Hutchinson. Before we can understand what they did, however, we must first discuss how Gause came to appreciate the principle for which he is best known.

The Niche

The competition studies immediately suggested the general principle that no two species coexisting in nature occupied exactly the same niche. This idea was not new: it was present in Darwin's *Origin of Species*, where he used it to account for the extinction of intermediate and transitional forms, and to explain why an organism's range was often more restricted than that dictated by physical conditions alone. Karl Nägeli, an eminent European botanist of the nineteenth century, wrote in 1874 on the displacement of plant forms by their competitors in response to some questions raised by Darwin's work.[29] His argument included some mathematical modeling and was therefore relegated to the mathematics and physics section of the Munich Academy of Science, where it lay mostly ignored until Gause cited it in 1934. Competitive exclusion was also a central principle in non-Darwinian treatments of competition, as, for example, that put forth by Frederic Clements. Hutchinson has uncovered several statements of the principle in the late nineteenth and early twentieth centures.[30] The connection of the principle with the niche concept was first made in 1910 by Roswell H. Johnson, a geneticist. Joseph Grinnell, a natural historian, had stated the principle of separate habitats without the niche concept as early as 1904, introduced the niche himself in 1913, and by 1917 stated that it was axiomatic that "no two species regularly established in a single fauna have precisely the same niche relationships."[31] J. B. S. Haldane stated the principle in 1924 in the context of his mathematical discussions of selection, but scarcely anyone took note of it.[32] The idea of competitive exclusion was regarded by all as obvious and not particularly interesting. Gause himself considered his statement of the principle in 1934 to be a natural extension of the conclusions of Volterra, Lotka, and Haldane.[33]

By Gause's time it was natural to link the principle with the niche concept, which had been recently popularized in ecology by Charles Elton in his widely read *Animal Ecology*, published in 1927. But in connecting competitive exclusion to the niche, Gause actually made a significant departure from Elton's use of the concept in his book. Before Elton, the niche was a nontechnical term signifying an abstract space in the environment which could be full or empty. Elton gave it a more precise and less obvious definition which served to focus attention on food relationships within the community. The niche became the animal's position in the community, in the sense of its place in the food chain. The idea was in direct analogy to the human community, as he explained: "When the ecologist says 'there goes a badger'; he should include in his thoughts some definite idea of the animal's place in the community to which it belongs, just as if he had said 'there goes the vicar.'"[34] Although Elton discussed the niche mainly in terms of food chains, he did admit that other factors might be necessary for a more exact definition. He considered it to be a smaller subdivision of the traditional groupings of herbivore, carnivore, insectivore, and so on, which helped to underline the basic similarities between communities appearing quite different at first glance. As a general economic category, therefore, a single niche could be occupied by different species; for example, the many species of insects which fed on pollen and visited flowers in a community all belonged to the same niche. (This grouping would now be called a guild.)

This use of the niche concept did not include the competitive exclusion idea, although Elton later used the niche in the earlier sense of Grinnell, and in that instance it did imply competitive exclusion. He had observed that the number of species in an association was rather limited and suggested that it had something to do with the limited number of niches available to herbivores. His evidence came from observations of Grinnell and T. I. Storer that different species of ground squirrel in the Rockies occupied different niches and thereby avoided competing with each other.[35] But again, Elton drew no particular attention to the principle of exclusion by competition. Nor did Gause at first, but by 1939 he had come to see his principle in an entirely new light.

The change occurred after he had had time to reflect on his experiments in the light of larger questions in ecological theory. Russian ecology, under the leadership of V. V. Stanchinskii, was just starting to break free of the earlier "structuralist" school, which interpreted the structure of ecological communities as the result of certain fixed laws of organization. This point of view resembled Frederic Clements's notion that the community as "complex organism" was governed by laws of development just as the individual organism was governed. Though Clements had considered experimental studies of such processess as competition important, this school of ecology was essentially descriptive, with emphasis on the classification of communities. Russian ecologists, Gause among them, wanted to supplant what they considered to be a static, morphological approach to the community with a dynamic analysis based on the study of the relations between the species in the community. Though the communities were still perceived as definite entities with both organization and self-regulatory properties, the organismic viewpoint was firmly rejected in favor of an explanation framed in the language of dialectical materialism.

The central questions were those which had troubled previous generations of ecologists: What principles determined the number of species in an association? Why were some species common and others rare? How could the stability of the community be understood in terms of the relations among the populations in the community? In the summers of 1935 and 1936, Gause took his ideas to the field and conducted a series of experiments on the colonization of glass plates by microscopic organisms.[36] In the light of these experiments, he began to realize that the competitive exclusion principle, at first introduced just in passing, threw some light on these larger questions. In 1939 he restated his principle with more force:

> It has been demonstrated by mathematicians, and strikingly confirmed by experiments with infusorians, that when two or more species have slightly different modes of living, or, in other words, when they occupy different ecological habitat niches, they will live together indefinitely in essentially stable equilibrium. What is more important is the fact that only a certain definite combination between the concentrations of the species living together possess the property of maintaining stability. Any deviation from this stable combination will automatically, owing to continuous process of competition, lead to the reestablishment of the steady state. It seems that the central ecological problem of regulation in the composition of complex biotic communities is in these analytical investigations reduced to simplest

> terms. In the light of all this evidence one may claim that if two or more nearly related species live in the field in a stable association, these species certainly possess different ecological niches.[37]

Competitive exclusion was no longer just a laboratory curiosity, but the key to the structure of whole communities, whose forms evolved from the dynamic processes occurring between their constituent populations. From his experimental research, Gause identified the dynamic processes as competition and mutual aid (symbiosis). The competitive principle that no two species could occupy the same niche was therefore the explanation of community structure, in that the best-adapted species were those that occupied the principal available niches. This meant that "at the basis of the structure of the biocoenosis lies the 'niche' structure."

Regulation of species composition was achieved by the ongoing process of competition, which automatically readjusted the balance between species whenever it was disturbed. Charles Elton had also believed that niches gave the community its special structure and appearance, just as trades and professions might in a human community, but he was thinking mainly of food chains, not competition. With Gause's interpretation of the niche as a unit structure over which species fought for possession, its dimensions shrank immediately to fit only one species. The niche became a place uniquely belonging to a given species, a place where it alone could enjoy full advantage as a competitor.

Gause's perception of community structure made such ideas as Clements's "complex organism" unnecessary, and Gause explicitly rejected organismic metaphors. But his explanation was fully compatible with the viewpoint of dialectical materialism: this gave it a strong advantage at a time when Russian scientists were pressed to adopt dialectical reasoning. Gause could explain the rise of the structured community through the dynamic relations between the individual and the environment, expressed through the competitive relations between species; a fine example of dialectics in nature.

In drawing attention to the competitive exclusion principle, Gause underlined the idea that the function of ecological research was to account for the spatial and temporal relations of the community. The principle itself became prominent in ecology, first as "Gause's hypothesis," "Gause's axiom," and "Gause's postulate," and later as the principle of competitive exclusion, as it was renamed by Garrett Hardin in 1960.[38] Though Gause did not develop the use of his principle in any detail in the field, its potentially tautological structure caused problems to arise almost immediately when it began to be used widely in ecology and still prompts occasional protests that the principle is not useful.

The problem occurs when one tries to define the niche in terms of the ecological requirements of species. To say that no two species can occupy the same niche seems the same as saying that no two species have identical requirements. This in turn seems trivial. But the nature of the niche, that is, the exact nature of those ecological requirements, is not known in advance. What the principle does is to focus one's attention on the fact that species must find ways to partition limited resources in order to coexist. Though Gause himself did not pursue these problems having to do with the meaning of the niche, his work stimulated Hutchinson and his student Robert MacArthur to look more closely at ecological relationships to see how this partitioning had been achieved. From this principle they were able to elaborate a research program of remarkable scope, the details of which will be discussed in chapter eight. They used the principle not as an end in itself, but as a means of uncovering subtleties in nature which the ecologist might not think to look for.

The principle of competitive exclusion was actually independent of the logistic model of competition from which it was derived. It could have been enunciated from experimental results alone. But Gause did not conclude from this that mathematical models were superfluous to ecology. Certainly it was possible to dispense with models for the analysis of particular problems. But the larger the problem, the more necessary it became to use models to uncover principles and to build general theories. Gause knew that mathematical and experimental work alone could never lead to perfect accuracy in prediction. But as he explained, the motive for using these mathematical models was not to predict specific events; it was to explain the principles governing community structure. The model merely helped to make the principle discernible. In a complex natural setting, he felt, such principles would be exceedingly hard to uncover.[39]

These remarks, addressed to an American audience, were not without relevance to the changing climate of Russia as well. Ecology in general was in a precarious position in Russia by the mid-thirties, as ideas regarding science and political ideology shifted ominously. Survival in science demanded a certain flexibility. Ecology had been able to adapt easily enough to two changes. It had adopted the dialectical reasoning which was now deemed to be the proper method for science and for the interpretation of nature. Second, it had dropped the earlier analogies drawn between human and natural communities, as well as any attempts to include man within these biological spheres. Any hopes engendered in the 1920s of creating a human biology along the lines of Raymond Pearl had to be quickly shelved. As Alpatov wrote to Pearl in 1940: "It would be wonderful to do here some work in human biology but unfortunately homo sapiens is considered here somewhat out of the field of biology."[40] A third shift had unfortunate consequences for some ecologists who criticized government policies aimed at agricultural reforms. During the Stalinist period, criticisms of these large-scale reform projects could spell the end of a career. Stanchinskii's was in ruins by the mid-thirties as a result of his disagreement with government policies. Finally, these years were marked by increasing opposition to mathematics in biology, especially by T. D. Lysenko and I. I. Prezent, both ill-equipped to understand any mathematical arguments. Gause, Severtzov, and Alpatov were criticized in 1939 for their mathematical work, which was thought to show an unpatriotic mimicry of foreign views.[41] Although Gause had indicated that he hoped eventually to combine the methods of Volterra with the mathematical genetics of Haldane and others, he did not follow up this plan of synthesis.

Gause and Alpatov had both moved away from mathematical studies and were now researching experimental problems in evolution from a nonmathematical point of view.[42] I do not know whether the criticisms directly motivated this change in research; however, the change was clearly a prudent one. In the late 1930s they began a series of investigations on the adaptability of protozoa to various concentrations of salt and quinine. These studies were related to a number of researches begun by Russian biologists in the mid-thirties concerning the ways in which acquired modifications might be fixed genotypically under the influence of natural selection. This theory, known as "organic" or "coincident" selection, postulated that certain modifications could be substituted by coincident mutations, provided that the mutations were associated with a selective advantage. The explanations were therefore not Lamarckian, but wholly in keeping with natural selection and mutation theory.

Though explored independently by Russian biologists, the concept of coincident selection had been popular in America and England at the turn of the century, just prior to the rediscovery of Mendel's work. The principal scientists responsible for developing the idea were James Mark Baldwin, Lloyd Morgan, and Henry Fairfield Osborn. Debate on its merits lasted from about 1896 to 1905 before Mendelism overshadowed the problem and discussion came to a halt. In the 1940s and 1950s, probably under the stimulus of the Russian research, the theory was also revived in the West and attracted some attention from J. S. Huxley, G. G. Simpson, and C. H. Waddington.[43]

Gause developed the theory in collaboration with Alpatov during the 1940s, publishing a detailed summary of his work in English in 1947.[44] During this period he also became interested in medical applications of his findings. These were an extension of the general problem of the struggle for existence, the first article in this area being published in Russian in 1943 with the colorful title, "The Struggle for Existence in Microorganisms in the Service of Wound Healing." Apart from a brief excursion into the problem of optical activity and living matter,[45] his researches took on a more practical tone in the 1940s. In 1949 he published a book, *Lectures on*

Antibiotics, and thereafter was wholly involved with antibiotics and microbiology, with special emphasis on the effects of carcinogenic agents on microorganisms. In the 1960s he published two books in English on his research: *The Search for New Antibiotics* (1960) and *Microbial Models of Cancer Cells* (1966).

The transition from ecological to medical research appeared from the literature to be more abrupt than it actually was, but in view of its close coincidence with the rise of Lysenkoism, it raises questions about the possible impact of Lysenko on Gause's research. His evolutionary studies provided a reasonable scientific alternative to Lysenko's doctrine of the inheritance of acquired characteristics, but they would have conflicted with Lysenko's position because they accepted the validity of evolution by mutation. There were also similarities between Gause's research and that of the evolutionist Ivan Schmalhausen, who was repeatedly denounced during the crucial conference of 31 July to 7 August 1948, which marked the start of the most intensely dogmatic Lysenkoist period.[46]

However, the University of Moscow, where Schmalhausen was located, was a center of anti-Lysenkoist sentiment and drew a great deal of fire from the conference participants, whereas Gause had left the university in 1941 and was working for the Academy of Medical Sciences since that time. But on 14 January 1949, the *U. S. News and World Report* announced that "C. F. Gause" [sic] was one of three scientists prominent in medical research who had been fired from key posts.[47] This rumor was repeated in two other discussions of the events of the time, first by R. C. Cook and from there by Julian Huxley.[48] The rumor does not appear to have been true, although there was decided pressure in Russian biology at the time to move into practical research. He won the Stalin Prize in 1946 and published steadily on antibiotic research in Russian and English journals throughout the most vigorous Lysenkoist period, 1948–1955.[49] In 1957 he visited America for the first time, lectured in New York on antibiotics, and afterward visited Hutchinson at Yale to discuss population ecology. In 1959 he again visited several cities in America and lectured in Hutchinson's laboratory. But by this time his ecological interests were peripheral. Although his later research was a natural outgrowth of his early interests in the struggle for existence, he had diverged considerably from the original context in which his work had first appeared.

Gause's ideas were very widely known, though his French monograph was much less known than his English language work. In general his ideas received much praise in the reviews. But being cited and praised is one thing; to have one's ideas put to use in the making of new science quite another. Gause's experiments were mentioned in the 1930s much more often than they were actually discussed or used. R. N. Chapman found them stimulating, but not immediately relevant to his own style of research. Thomas Park found them useful only in the most general sense that they focused attention on the neglected problems of competition, evolution, and speciation, but he did at least discuss them in some detail. Charles Elton, in his *Voles, Mice, and Lemmings* of 1942, found the need for scarcely a couple of references to Gause's work in nearly five hundred pages of text.[50]

The experiments were often cited to buttress a pitch for more experimental work, but they were not themselves made the basis of such research right away. One of the main problems was simply their laboratory setting. It seemed unlikely that the simple environment of the laboratory could have much to offer to the analysis of the more complex field situation. Competitive exclusion, often noted before Gause, had no special claims to being a universal principle. The experiments sat quietly in the literature, admired but on the whole neglected.

The neglect did not last. By the mid-forties Gause's principle was on everybody's lips, having been brought to prominence simultaneously by Hutchinson in America and David Lack in England. Though each had a different problem in mind, they both perceived that when trying to understand complex events, often the best way to start is with a very simple question. Lack's studies of evolution in the Galápagos finches will occupy the next section, leaving Hutchinson's rather different approach for the next chapter.

A "Heterogeneous Swarm" of Finches

David Lack's question concerned Darwin's finches and how closely species could resemble each other ecologically. These birds had been a puzzle ever since Darwin had studied them. Starting from a single ancestral form, they had diverged on the Galápagos Islands to fill niches that on the continent would have been taken by several other kinds of birds. The species differed in body size and plumage, but the most striking differences were in the size and shape of beak. One could find here "small finch-like beaks, huge finch-like beaks, parrot-like beaks, straight wood-boring beaks, decurved flower-probing beaks, slender warbler-like beaks; species which look very different and species which look closely similar."[51]

The species which looked very different were easy to interpret; they were obviously adapted to different stations and food habits. But there was no apparent explanation for the species which looked very similar. Of the seed-eating ground finches, for instance, three species differed only in size of body and relative sizes of the beak, but all had beaks of the same shape. They could be found in mixed flocks, all feeding together with no obvious ecological separation between them. Harry S. Swarth of the California Academy of Sciences studied them in 1932 and concluded that these variations in size were not correlated with variations in food.[52]

DAVID LAMBERT LACK, 1910–1973
Photograph from Alexander Library,
Edward Grey Institute of Field Ornithology, Oxford

Natural selection was therefore ruled out as the cause of these differences. He believed that as the birds had diversified on the islands, certain characters such as beak size had simply become "unstable." Presumably he meant by "instability" only that divergence had occurred without reference to adaptive value, and was not intending to suggest some genetic mechanism peculiar to the finches.

The case of the Galápagos finches was merely a recent flare-up of an older debate that had flashed with especial intensity in the late nineteenth century.[53] At that time it was George Romanes who was arguing that the differences between closely allied species were not adaptive and were therefore not the results of natural selection. Romanes's argument led to a distinction between the process of speciation, that is, the origin of minor systematic diversity, and the process of adaptation. Natural selection was thought to explain adaptation, but not the initial formation of distinct species from one parental form. This distinction between the origin of species and the origin of adaptation persisted into Lack's time. In opposition to Romanes, A. R. Wallace argued that apparently useless characters might well turn out to be useful upon closer examination. Among those who followed the staunch adaptationist position taken by Wallace in the early twentieth century, the existence of adaptation was often assumed without concrete, experimental proof. Observations of mammals, for instance, often showed that closely related species occupied separate habitats or niches. It was usual to assume that they were structurally or physiologically adapted to these different habitats.

But in the 1920s a reaction to the adaptationist position began among field naturalists. A. J. Nicholson's early work on mimicry had been addressed to the problem of how the usefulness of characters should be defined. His studies of butterflies suggested that a character arising through competition within the species might not be useful to the survival of the species as a whole.[54] In 1926 O. W. Richards and G. C. Robson argued that the differences between closely allied species were not adaptive and that the role of natural selection in producing them was "very dubious."[55] Theirs was the common systematist's point of view; systematists used these small, apparently nonadaptive differences to distinguish closely related species. Recent historical research by Provine, Gould, and Kimler has uncovered just how extensive the nonadaptationist view was among English and American naturalists up to the mid-thirties, though Gould suggests that the situation was different in Germany.[56] Provine points out that the argument of these naturalists and taxonomists was responsible for the interest taken in Sewall Wright's idea of random genetic drift as the most likely explanation for nonadaptive differences (although Wright's mature shifting-balance theory emphasized adaptive responses at the level of geographical races and closely allied species, downplaying the role of drift).

On the ecological side, Charles Elton concurred with Richards and Robson and suggested that chance mutations, neither harmful nor beneficial, could become established in populations while their densities were reduced and when natural selection was not acting.[57] J. B. S. Haldane argued against Elton's view on mathematical grounds in 1932, though in 1940 Wright still defended the idea, for it supported his own theory.[58] But in 1931 E. B. Ford had come up with an attractive compromise in his widely read *Mendelism and Evolution*.[59] Ford agreed with Richards and Robson that some characters were nonadaptive, but he disagreed that these characters were established by a process of nonadaptive evolutionary change. Useless or even nonviable characters, in order to persist, must have become modified through a kind of stabilizing selection, where the nonadaptive character was actually associated genetically with an adaptive change of a more subtle nature. His field studies on butterflies gave strong support to his argument.[60]

Ford's views reflected the influence of R. A. Fisher, who was most adamant in arguing against the "dogma" that many specific differences were not adaptive.[61] Fisher inferred that it was only from ignorance of the ecology and life history of populations that the adaptive value of these differences had not yet been understood. Julian Huxley also leaned to the adaptationist side, hoping in 1936 that the current skepticism would prove to be a passing fashion and that the organism would come to be seen as a "bundle of adaptations, more or less efficient, co-ordinated in greater or lesser degree."[62] But in 1936 the antiadaptationist position was still strong, though weakening, and the problem was the subject of much discussion at zoological meetings in England.[63]

The Galápagos finches provided an interesting testing ground for these views. Percy Lowe, ornithologist at the British Museum (Natural History), published an overview in 1936 in which he asserted again that there was "no scope for Natural Selection" when it came to this "heterogeneous swarm" of finches that was the despair of systematists.[64] He advised more on-the-spot observations and breeding experiments to help make sense of the bewildering diversity. Moreover, he reminded his readers, as Swarth had also done some years earlier, the finches were unusually willing subjects for a naturalist's study, for they were so tame that they could "almost be picked off the bushes."

Lack found Lowe's suggestion most attractive. He had been looking for a group of related species to do a comparative study of territory in birds; the Galápagos had much to offer. But the islands were very far away. He hesitated, until Julian Huxley came to the rescue. Lack had known Huxley since his graduation from Cambridge in 1933. It was Huxley who had steered him toward his job as biology teacher at the progressive Devonshire school, Dartington Hall. Since then Huxley had acted as Lack's unofficial supervisor, listening to his ornithological discoveries and "providing immense enthusiasm and stimulus, though not criticism." Now Huxley gave more than encouragement: he got grants from the Royal Society and the Zoological Society of London to finance the expedition. After a year's delay, Lack and his group set off in the autumn of 1938.[65]

He found the archipelago depressingly unlike the "Enchanted Isles" he expected from the travel books. The dreary scene of thorn scrub and lava was made nearly intolerable by a host of petty aggravations: "food deficiencies, water shortage, black rats, fleas, jiggers, ants, mosquitoes, scorpions, Ecuadorean Indians of doubtful honesty, and dejected, disillusioned European settlers."[66] The finches themselves, for all the excitement their great variety aroused, were dull to look at as they hopped about the ground "making dull unmusical noises." On the return voyage in the spring of 1939, the birds that Lack was bringing to London began to moult as the ship reached Panama. A detour was made to San Francisco to save the birds; there at the California Academy of Sciences was the largest museum collection of these finches. Lack stayed on for five months to study this collection and others housed in the United States before returning to England in the autumn. On his way home, he also took the opportunity to visit Ernst Mayr in New York for several weeks. He had met Mayr on a previous trip in 1935. They both got along well, being ornithologists and committed Darwinians. Mayr had become converted to natural selection from a Lamarckian stance, whereas Lack had always been a keen Darwinian, without its Lamarckian overtones, having been introduced to evolution at school through the books of W. P. Pycraft.[67] The friendship with Mayr would later have an important impact on Lack's thought.

Back in England, he quickly wrote up his results on the finches; then, as the war started, he became involved in work for the newly formed Operations Research Group. Part of his job in 1942 was to tour the ship-watching coastal radar stations. His companion was his old friend from his student days at Cambridge, George C. Varley, with whom he spent much time discussing population biology. Lack was just completing a book on natural history, *Life of the Robin*, and he decided to write another for his sixth-form pupils using the Galápagos finches to illustrate evolutionary ideas. As he thought again about these birds, he became curious to know how closely different species could resemble each other ecologically. This question led him, over a six-month period in 1943, to revise his earlier ideas, already written up but delayed in publication because of the war, on the evolution of the finch species.[68] In the course of revising his views, as he later recalled, he came to appreciate Gause's principle of competitive exclusion. He had read Gause's statement of the principle in 1939, but like many others assumed it could apply only to the special circumstances of the laboratory. In *The Struggle for Existence*, Gause had actually cited a field example, studied by A. N. Formosov, of species of terns living together with no apparent conflict because their methods of procuring food differed. Formosov used this material in his lectures in Moscow, but never felt he had enough information to publish.[69] However, at this stage Lack appears not to have been familiar with Gause's book, but only with the shorter statement of 1939.

From what he had understood in 1939, the Galápagos ground finches struck Lack as an obvious exception to the principle, for they seemed to occupy identical niches, yet they did not interbreed. But in 1943 his opinion changed, and the finches became instead paradigm cases of Gause's principle in action. The principle in turn became an important part of his new theory of evolutionary divergence and the role of competition in promoting such divergence. The didactic work for his pupils turned into a different kind of book entirely, a theoretical work in evolution called *Darwin's Finches*, which came out in 1947.

Several events may have helped Lack to change his mind about the finches. His conversations with Varley and the appearance in 1942 of two monumental studies in evolution—Huxley's *Evolution: The Modern Synthesis*, and Mayr's *Systematics and the Origin of Species*—all helped to clarify his thinking. The starting point was the problem of how populations became isolated as a prelude to speciation: this was a central question of the 1930s and 1940s. Huxley took a broad view, arguing for three kinds of isolating mechanisms: geographical, ecological, and genetic.[70] Geographical differentiation meant spatial separation, which would be followed by biological divergence. Genetical differentiation referred to a change in the genetic machinery of heredity, sex, and reproduction, which either prevented interbreeding or made hybrids less fertile. Ecological differentiation referred to a specialization in function. Huxley thought that ecological divergence would occur when there was little or no competition, but when different parts of a population were under different selection pressures, causing them to adjust to different modes of life. He believed that such functional specialization could lead to full speciation, even within a single geographical area. This emphasis on ecology came closest to Darwin's own image of how species diverged. These three mechanisms overlapped and combined, making it immensely difficult to arrive at a satisfactory definition of the species.

Like Huxley, Lack also emphasized ecological factors in evolution. His first ornithological idea had been the notion of habitat selection, the idea that divergence in birds had at times occurred through the selection of different habitats.[71] This explanation focused on psychological factors: a bird might choose a habitat simply because it preferred it, not for any reason connected to survival or to breeding limits. Further conversation with Huxley would certainly have reinforced Lack's appreciation of ecological factors in evolution, the classical Darwinian stance. But upon his return from America in 1939, Lack no longer believed that ecological isolation was likely to occur before speciation. Instead he had come to adopt Mayr's position that only geographical isolation could lead to speciation. For Mayr, ecological differences were only secondary products of spatial separation. This was a clear difference between him and Huxley. Mayr's arguments were most persuasive. Lack abandoned his habitat-selection theory, and Huxley, too, after reading *Systematics and the Origin of Species*, admitted that Mayr had caused him to change his own views of the relative importance of geographical and ecological isolation.[72]

As Lack was reading Mayr's book, emphasizing the primacy of geographical isolation, and Huxley's book, emphasizing the importance of ecological interactions, he was also debating with Varley the problems of population regulation. Varley had the opportunity to remind Lack of an article he had read as a student and promptly forgotten: Nicholson's theoretical study of competition and population control, published in 1933. Varley had fashioned his Ph.D. thesis around Nicholson's ideas and had made the first field test of a few of Nicholson's predictions.[73] When Lack and Varley got together later, Lack helped him to rewrite the thesis for publication. It appeared in 1947, over a decade after the research had been done. The delay was caused by the fact that Varley had first sent it to the Royal Society for publication. It was referred, as he knew it would be, to W. R. Thompson for comments. Predictably, it came back heavily criticized.[74] Varley finally published it in the *Journal of Animal Ecology.*

Varley and Lack had discussed the question of when competition would be important for populations living together. Varley pointed out that if numbers were controlled by parasites or predators, the populations might be kept at a low enough level to avoid competition. In that case, two species might well occupy the same niche.[75] (Gause was perfectly aware of this possibility and had given it fuller treatment in his French monograph, which few people seemed to have read. It remained for Hutchinson to bring this book out of oblivion in the 1950s.) If food supply was limiting, however, Gause's principle suggested that competitive exclusion must occur. This was closer to the situation on the Galápagos Islands. The logic behind Gause's argument now struck Lack as inescapable. Two populations might diverge when geographically separated, but it was still necessary to explain how they persisted once they met again in the same area. A full evolutionary explanation had to combine the problem of isolation with the problem of continued survival and ecological adaptation. Surely the species could persist only if they occupied separate niches. Lack took another look and this time saw the ground finches quite differently. It had been noticed ever since 1902 and confirmed often afterward, that although the finches ate the same kinds of food, there were small differences in the size of food they took. Birds with larger beaks tended to eat larger food. Lack and everybody before him had thought these differences were incidental; but now, in the light of Gause's hypothesis, he considered them to be significant. By differing ever so slightly in their food habits, these species were able to reduce competition among themselves. Therefore, the differences in the beak sizes were adaptive. Huxley had also suggested in 1942 that species in the same area might have evolved size differences to reduce competition, but at the time Lack had paid no attention. Now, as he described it, he "painfully rediscovered" the idea for himself.[76]

Lack was convinced that these species had originally diverged at a time when they were geographically isolated. When they met again, they had reduced the competition between them by becoming ecologically specialized and were able to coexist. The differences between closely allied species were not trivial and meaningless, as Richards and Robson had argued; they were adaptations, mechanisms to avoid competition. The origin of minor systematic diversity was not a problem separate from adaptive specialization, as most people had thought; rather, Lack suggested, it was an agent promoting specialization. Gause's principle was not a trivial statement about laboratory populations; it could help in the understanding of evolutionary puzzles in the field. The finches, far from being exceptions to the rule, were perfect examples of the principle at work. And the exclusion principle provided something badly needed in ecology, a simple hypothesis which could be used to look at ecological processes and bring some order to this complex subject.[77]

Lack was not yet willing to extend his ideas universally. He still thought many kinds of subspecific differences had no adaptive value. But at least in some cases, competitive exclusion might account for differences that had ecological significance. His argument, he realized, was based largely on inference and circumstantial evidence; he had not proven these inferences experimentally. Nor could he imagine an easy way to demonstrate the truth of his carefully constructed argument, for competition among mobile creatures was hard to test. It was easy enough to show competition among plants by transplanting a species to another habitat and watching it be eliminated. A bird would simply fly away.[78] Instead of direct proof, Lack relied on deduction from two likely premises: that divergence arose from geographical isolation alone and that the species were limited by food supply. The facts which he brought to bear on his argument supported his interpretation, but it was not a clinching argument.

Lack was fully aware of the circumstantial nature of his argument. In his writings, however, he adopted a position of strong advocacy, a style suggestive of the didactic aims of the schoolteacher. His first opportunity to express his views came at the meeting of the British Ecological Society in 1944. The subject chosen for discussion was the ecology of closely allied species, with Gause's hypothesis at the focal point. Elton, Lack, and Varley all argued in support of Gause's principle, while three others attacked it.[79] Cyril Diver's vigorous criticism expressed what had by then become a common sentiment, that mathematical and experimental approaches had been "dangerously over-simplified and omitted consideration of many factors, of which the importance varied among different organisms." His point was well taken, for although the principle itself seemed logically necessary, in practice its application required knowledge still barely available on the factors limiting populations and on the importance of competition. For Lack these difficulties did not matter. The important thing was that the principle provided a way to get at new knowledge. His method was to formulate a hypothesis, then to look for its verification in the field. He saw where the rule fitted; others saw only the exceptions.

When Lack published *Darwin's Finches*, he felt his views to be in the minority. When the book was reissued in 1961, he could report with satisfaction that the views presented there about the prevalence of adaptation had come to be widely accepted. Coming at the end of the evolutionary synthesis, his book reflected the trend of the times—a growing emphasis on natural selection as the explanation of evolutionary diversity. Stephen Gould has referred to the strong adaptationist position which developed from the 1930s to the 1950s as the "hardening of the synthesis." Lack's argument was successful not only because of its strong Darwinian bias, but because of the way it was presented, with the strength of absolute conviction. When a problem could be seen from two sides, as this one could, it helped to make a clear assertion of faith. Given the alternatives, Lack reasoned toward his position and stayed with it. And if something is said positively and often enough, it will first generate debate, then generate experiments, then perhaps win adherents.

Indeed, his book established itself as an authoritative study so quickly that Lack has been singled out as the main source of our modern acceptance of the idea that these finches inspired Darwin himself to his evolutionary views and gave him insight into geographical isolation and adaptive radiation. Frank Sulloway has argued, to the contrary, that this image of Darwin and his finches is comparable to Newton and the apple.[80] In fact, Darwin did not understand the geographical distribution of the finches, did not appreciate the importance of geographical isolation for speciation, and did not even mention the finches in the *Origin*, although he referred to the Galápagos Islands six times. The evolution of this myth started with studies of the Galápagos fauna after Darwin, which brought the finches into greater prominence, until Lack's book resolved the long-standing debates on these birds. But Lack also presented his study as a triumph of Darwinian theory, and in doing so he gave Darwin more credit for understanding the significance of the finches than he deserved. Darwin simply did not know enough about the distribution of the species to have seen the finches in the way Lack suggested.

Be that as it may, Lack's argument has been a continuing stimulus to research. The questions he raised are still being investigated, elaborated, and now challenged, based on studies of the Galápagos finches.[81] Gause's hypothesis also has become a source of continuing conflicts over the interpretation of competition as a determinant of community structure.[82]

Years of Appraisal: The Fifties

By the time *Darwin's Finches* was published, Lack had gladly given up teaching school to become a professional ornithologist. In 1945 he became the director of the Edward Grey Institute of Field Ornithology, which had been running on a shoestring budget at Oxford since 1938. The institute's position improved by 1948 when it was joined with Elton's Bureau of

Animal Population to form a new Department of Zoological Field Studies, though the two halves operated independently of each other. Population ecology in general seemed finally to be established as a distinct branch of ecology. In 1946 Thomas Park declared its permanence.[83]

At Elton's laboratory, Leslie continued to work closely with ecologists and to introduce them to demography and modeling techniques. In 1947, L. Charles Birch, an Australian entomologist, visited the bureau and with Leslie's instruction began to use demographic methods in his studies of insect populations. He had come to Oxford from Park's laboratory in Chicago and the next year was joined by Park himself, who also became involved in intensive theoretical work with Leslie. Park's earlier work on competition among flour beetles was not mathematical, but it was intended to unravel the basic mechanisms of competitive interaction. His conclusions were summed up in his distinction between "exploitation" and "interference" competition, terms which later gained wide currency.[84] Interference referred to the direct interference between individuals in a population, whereas exploitation occurred when two species were exploiting a resource present in limited supply. In the 1950s, Park teamed up with Jerzy Neyman and Elizabeth Scott to see whether mathematical models could be devised for the beetle populations.

Lack meanwhile continued his work in evolutionary ecology, much of it plunging him into various controversies over the importance of competition and the interpretation of natural selection.[85] In the 1940s and 1950s, he took up the problem of the evolution of clutch size in birds, using as a hypothesis the novel idea that clutch size would evolve by natural selection to produce the greatest number of young surviving to independence. This meant that clutch size would not be adjusted to physiological capability, nor to mortality, but to the food supply available to the parents to feed their young. His work on the evolution of reproductive rates was an extension of Nicholson's insight that, in order to understand fitness, it was necessary to know how populations were regulated. Lack's study of these problems was published as *The Natural Regulation of Animal Numbers* in 1954, a book which owed much to his conversations with Varley.[86]

Some of his ideas sparked disagreements with L. C. Birch and his colleague H. G. Andrewartha, whose own point of view was developed in *The Distribution and Abundance of Animals*, also appearing in 1954.[87] The two books were different in their goals, with Lack's addressed to problems of evolutionary strategy, while Andrewartha and Birch were more concerned with the daily events of the field. Lack chose to focus on ultimate causes, or those concerned with ultimate survival value, rather than the proximate factors which dealt with specific physiological and behavioral responses to environmental cues. The distinction between these two types of explanation had been brought to his attention by John R.

Baker in 1938.[88] Andrewartha and Birch, on the other hand, were interested in proximate causes. This did not mean they were oblivious to evolution. Birch was still involved in experimental studies in evolutionary biology, and their book did include chapters on genetic changes and evolution, though these sections were slim compared with the rest of their massive volume. They saw evolution as a continuous process of change, however, as something that could best be understood through the study of the daily processes of an animal's life. Their concern with the details of these processes caused them to be skeptical of mathematical modeling, though mathematics in the form of demography was acceptable.

They were equally suspicious of claims about the importance of competition, for it was not clear that competition was one of the ongoing processes of daily life. Lack believed that the circumstantial evidence combined with logical necessity made the argument for competition among birds compelling, but he was trying to account for the evolution of adaptations at some time in the past. In the insect populations which Andrewartha and Birch studied, it seemed that the explanation of abundance could be adequately presented by referring to known effects, such as climate. Birch even disliked the very word "competition" because of its implications of emotion or purpose.[89]

This dispute was only a small part of a general malaise seeping through population ecology as different approaches were brought face to face, each claiming its own worth. The postwar period through the 1950s saw the reappraisal and elaboration of many of the theoretical ideas to date, with the call for more experiments, first made in the 1930s, finally being heard. Gause's experiments were extended chiefly by an Australian then working in Cambridge, Alistair C. Crombie, who shortly after gave the whole thing up to become a historian of medieval science.[90] The Lotka-Volterra predation model became the focus of laboratory tests, as did Nicholson's ideas.[91] There was greater collaboration between biologists and mathematicians, resulting in the exploration of more realistic stochastic models.[92] These collaborations helped to instruct biologists about the nature of these models, whose properties were imperfectly understood.

The stochastic version of the Lotka-Volterra equations, for instance, discussed by Maurice S. Bartlett in 1960, showed that an unstable condition would prevail where the extinction of either the predator or of both species was more likely to occur.[93] Nicholson and Bailey noted the relevance of this result to their own findings of instability using discrete-time models.[94] The relationship between all these models was still puzzling, but it was clear that modifications of the original deterministic models would be necessary, as W. R. Thompson had cautioned years before. There also needed to be more care in experimental design, for as Frederick E. Smith noted in 1952, "Almost all of the evidence used in support of various interpretations is, at best, inconclusive."[95]

With more information available and a better grasp of statistical analysis, the fires of discord which had flashed in previous decades were not damped. On the contrary, more knowledge provided more fuel. The debates continued, revolving around the problem of how populations were regulated, whether by density-dependent or density-independent factors. A scan of the voluminous literature reveals that the conflicts were operating on several levels at once, reflecting the eclectic nature of the community. Population biologists were deeply divided, a "heterogeneous unstable population," as Hutchinson called it in 1957 when ecologists and others in the population field gathered at Cold Spring Harbor for a symposium on animal ecology and demography.[96]

One level of disagreement concerned the interpretation of the evidence: What indeed *were* the facts? Lamont C. Cole, a student of Thomas Park, took on the question of population cycles and suggested they might simply be random series.[97] He argued in 1951 that the possibility of random cycles ought to be eliminated before more complicated explanations were invoked, though an analysis of the lynx cycle in Canada being done at the same time by P. A. P. Moran indicated that this cycle at least was real.[98] The study of competition was beset by similar problems, for competition was assumed more often than it was demonstrated.

Laboratory experiments were intended to resolve this problem of what was real and what illusory. Experiments could not mimic field conditions, but they could serve as models or illustrations of theories and could help to unravel the biology of the situation. Nevertheless, there were still strong feelings that experiments would not reveal much about field conditions in the long run and that they were being used to substantiate mathematical theories which were clearly wrong. The debates on the relation of theory to practice, on the validity of mathematical models, leaned one way or another depending on what kind of population was studied—human, mammal, bird, or insect. In the laboratory or in the field, different points of view, different techniques and assumptions, inevitably gave rise to conflicting views about which methods and which questions were appropriate.

Much of the controversy centered on the definition of basic concepts. To an outsider the whole argument seemed to be mainly about words.[99] The understanding of density-dependent and density-independent regulation was inconsistent, as was the whole notion of regulation. The very idea of the population itself was unclear. W. R. Thompson had argued that the population existed only in the mind and had no objective reality. Consequently, any discussion of self-regulation within a population was "merely playing with words."[100] Nicholson, who had observed natural selection in his laboratory populations over the years, was finally moved to define the population along the lines of the population geneticists: "a group of interacting and inter-breeding individuals which normally has no contact with other groups of the same species."[101] He also tried to throw some light on the concept of competition by distinguishing between competition in territorial and nonterritorial species as "contests" and "scrambles," respectively.

All of these levels of disagreement were extensions of the same problems that had dogged population ecology for the previous three decades since the introduction of mathematical models. Exacerbating these very real problems of how nature was to be interpreted were additional confusions caused by a tendency to search for single causes as explanations, in spite of the frequent caveats against doing so. There was, however, a new development in the 1950s which was creating important changes in population ecology and was impinging upon these old debates. The change was a greater interest in evolutionary problems, following the stimulus to neo-Darwinian theory provided by the evolutionary synthesis of the mid-thirties to mid-forties. Ecologists had been aware of evolutionary problems well before the synthesis, and some had argued continually for better integration of ecology and evolution. But they felt themselves to be in a minority. Ecology's main purpose was still to understand organism-environment relations and the principles of community structure and development. These traditional aims continued to be important to ecologists in the 1950s, but now they were joined by a more explicit interest in evolution.[102]

The concern with evolution stimulated interest in the demographic approach pioneered by Pearl, Lotka, and Leslie, and in genetics by Fisher. Cole pointed out in 1954 that knowledge of demography was necessary because natural selection would be expected to shape life-history patterns to conform to efficient populations.[103] His article helped to explain to ecologists how to think in "strategic" terms, that is, to see the entire life history of an animal as a strategy designed to maximize fitness, however fitness might be defined. The use of "strategic thinking," which would help to bring ecology and genetics closer together in the 1960s, had been anticipated much earlier by Lotka, though in a nonevolutionary context. Lotka freely borrowed from related problems in economic theory, military strategy, and game theory. Now in the 1950s ecologists independently began to pick up mathematical techniques from these other fields as well. The modern version of the nineteenth-century "argument from design" projected onto nature not God's design, but man's. Concern with evolutionary strategies and principles of optimization of design were aspects of the wider interest in adaptation which followed the modern synthesis and of which David Lack's work was a part. Gould, in identifying the "hardening" of the synthesis, admits he has no general explanation which fully accounts for this shift. It may be more than coincidence, however, that a hard adaptationist line was associated with a mathematical, engineering point of view which emerged shortly after the war.

In any case, there was by the 1960s a new schism in ecology, as perceived by ecologists at the time, between those interested in ultimate causes and those interested in proximate causes. (This distinction had been recently popularized by Ernst Mayr.)[104] What is significant is that the evolutionary approach was seen to be better because it promised a unifying ground for ecology. Gordon Orians, contrasting Lack's style with that of Andrewartha and Birch, concluded that only natural selection could provide a general theory of ecology. Andrewartha and Birch's approach was made to seem less important than the burgeoning areas of theoretical ecology explicitly addressed to evolutionary problems.[105]

Lack's style was not mathematical, but the idea that ecology should move in the direction of evolutionary theory naturally raised questions about the role mathematical models could play. In 1957 Lack was visited in Oxford by a young man from America who was impatient to set ecology on its mathematical, evolutionary track. Robert MacArthur was exasperated by the arguments of the previous years. To be sure, some populations, like the ones Andrewartha and Birch studied, fluctuated widely from year to year in response to unpredictable changes in the environment. But other populations, like the birds Lack studied, were fairly stable. If mathematical models could not be applied easily to the unpredictable populations, there seemed no reason to be as wary of using mathematics in equilibrium populations. And where models could be used, a chance of creating a new theoretical ecology also existed.

This was a risky undertaking, for it meant turning away from a vast amount of ecological research in order not to be distracted by too much information. It meant being a little bold in defining what was important to ecology, knowing that many ecologists, wrapped up in long and detailed studies of particular organisms, would find the use of mathematics far too abstract and irrelevant to their concerns. MacArthur grasped the nettle.

8

The Eclipse of History

Robert MacArthur liked to make predictions. Reviewing the Cold Spring Harbor Symposium of 1957—which was notable, as he said, more for its "almost religious fervor" than for any new facts—he briskly dismissed it as one that would soon be forgotten.[1] He was wrong. The reason he was wrong has a great deal to do with the concluding remarks to the symposium made by Evelyn Hutchinson.[2] These were not the sort of concluding remarks one might expect at such a meeting, a summarizing account of "what we have learned." What Hutchinson offered was a long meditation, a weaving together of theory and observation, shifting from the abstract to the concrete, compounded "of equal parts of the obvious and the obscure." It was also a weaving together of his ideas and those of his student, MacArthur.

His own preoccupations centered on the definition of the niche concept; MacArthur's ideas concerned a model of species abundance introduced the year before. Both the definition and the model were designed to investigate how species lived together in theory and in practice. The discussion revolved around Gause's hypothesis. Many ecologists had been inclined to reject the hypothesis as trivially true, but Hutchinson hoped to show that it could be used to ask meaningful questions of nature. Moreover, he hoped to divert attention away from the fruitless arguments so evident at the meeting and toward some broader themes of ecology and evolution. Contained in these remarks were the germs of an entire program of ecology yet to unfold. The program was mathematical; it focused on competition as the key to understanding community structure; and it stressed the need to integrate fully ecology with evolutionary biology. It was a direct outgrowth of the mathematical ecology of the previous decades, channeled largely through the work of Gause, but pushed well beyond Gause in its applications to various branches of ecology.

Hutchinson and MacArthur helped to make ecology intellectually exciting. In retrospect, it is easy to find weak spots in specific ideas or methods, yet together they were successful in bringing mathematical methods which had been lingering on the fringes of ecology squarely into the mainstream of this science. The fact that their methods and assumptions are now being criticized with some vehemence says as much about

George Evelyn Hutchinson, b. 1903
Photograph by William K. Sacco

their pervasive influence as it does about any weaknesses. Some of these attacks have a polemical cast which has created a mood of intense animosity hardly different from the fierce encounters I have already described. I shall not wade very deeply into this modern controversy, though I shall indicate as I go along some of the sources of conflict and the relevant literature dealing with it. Here I am mainly concerned to describe what Hutchinson and MacArthur did, to show why their work was felt to be so exciting at the time.

The Paradox of the Plankton

Hutchinson was educated at Cambridge and taught zoology for two years at the University of Witwatersrand in Johannesburg. This was a generally unhappy time as far as teaching duties went, but it brought the ultimate benefit of increasing his interest in fresh-water ecology.[3] In 1928 he crossed the Atlantic to start a new appointment at Yale University. In response to the work of V. I. Vernadsky and Viktor Goldschmidt in the 1930s, his interests by the 1940s were turning to biogeochemistry.[4] From his early days at Yale he had been kept up to date on Pearl's progress through Oscar Richards, who was writing his dissertation on logistic theory at the time. But it was not until Gause's later work that he began to see how the new arguments entering ecology might be put to work.

It all began with the paradox of the plankton. The problem was to explain the coexistence of different species of planktonic organisms in the homogeneous waters of the lake, which apparently constituted but a single niche. Hutchinson was first made aware of this paradox in an early study of successional changes in aquatic species.[5] It was not clear that these changes were directly related to chemical changes in the water. Gause's work suggested that making correlations between populations and environment was too simple an approach. It was more likely that changes in the chemical environment were affecting the dominance relations among the competing species and that these changing competitive relations were directly affecting the sequence of succession. Previous limnological studies had tended to stress the direct transgression of an organism's physiological limits of tolerance by environmental changes. But it now seemed that more complicated competitive interactions also played a part in the full story.

The idea that competitive relations were important in succession was not new, for Frederic Clements had also built his studies of plant succession on the foundation of competition. But Hutchinson soon elaborated his example into a more general methodological argument. The puzzle of this little paradox, first discussed in 1941 and refined over the next couple of decades, set the theme that would recur in all of his population writings: that a theory, even if it was tautological, could be used to investigate cases where its conclusions seemed to be false.[6]

These views were an extension of the philosophy of science developed by Harold Jeffreys in his book *Scientific Inference*, published in 1931. Jeffreys's work in cosmogony and geophysics had made him aware of the problem of applying physical laws beyond their original range of verification in time and distance.[7] Hutchinson, who knew of Jeffreys through his

Cambridge connections, saw the parallel between planetary physics and ecology. He read the book immediately when it was published, though it was not widely read by biologists.[8] It proved to be just what he needed.

Jeffreys started the book with a discussion of the place of syllogism in science and of the different ways of deriving the premises of logical arguments by deduction or induction. His view that the syllogism was an important guide to thought became in turn the core of Hutchinson's scientific method. Jeffreys admitted that many types of logical argument merely told one what was already known and therefore were not useful for science. But in other cases the premises of an argument might not be known with certainty and could be subject to falsification through experience. An inference found to be incorrect in this way could lead to an adjustment in the original statement. Making a scientific inference therefore was to make a statement of probability based on a degree of belief: "When we make a scientific generalization we do not assert the generalization and its consequences with certainty; we assert that they have a high degree of probability on the knowledge available to us at the time, and that this probability may be modified by additional knowledge."[9] The method of inference as he outlined it was intended to help assimilate new knowledge and to indicate how scientific laws might be improved. It was also intended to serve as a guide for the descriptive and classificatory parts of science.

Armed with Jeffreys's ideas, Hutchinson could take a more positive approach to mathematical arguments, such as the logistic curve and Gause's hypothesis, than many other people. Whereas the majority wanted to discard these imperfect theories, he preferred to tinker with them on the chance that they might lead to something which would work. Though he was not a mathematician, his training was good enough to enable him to apply mathematics to biological problems with little difficulty. In a series of review articles and addresses starting in the 1940s, he did exactly that, sorting piece by piece through the copious literature of field research to see what observations could be related to theoretical ideas.[10] The first piece was a review of Elton's *Voles, Mice, and Lemmings*, whose title, "Nati sunt mures, et facta est confusio," ("Mice are born, and the result is confusion") was both a comment on nature and on the style of ecology he felt the book represented.[11] He took care to compare Elton's observations with the predictions of Lotka, Volterra, and Gause, and he closed the review with the wish that Elton might make "a speedy return to his interrupted studies, so that, after many mice have been born, we may, in ten or twenty years, have a companion volume in which the confusion that now permeates the whole subject will be resolved into order." In other articles he tinkered with the models themselves. To the logistic equation he added a time lag, showing how this led to oscillations in the single population.[12] To the competition equations he added higher order terms to describe the competition resulting from social interactions.[13] All of these discussions helped to place these theories solidly in front of his ecological audience.

His students followed suit. A rough idea of what Hutchinson taught in class may be gleaned from a text published by Lawrence B. Slobodkin in 1961.[14] Slobodkin finished his doctoral work under Hutchinson in 1951, then took a post at the University of Michigan, where another former student, Frederick E. Smith, also worked. His book covered problems in niche theory, demography, logistic theory, competition, and predator-prey relations; and ended with a projection of a unified ecological theory which would provide answers to practical problems as well as unite both ecology and genetics into a predictive evolutionary theory. It did a great deal to spread the mathematical message of the Hutchinson school, even though Slobodkin himself later became more skeptical of mathematics. There was no single problem that united Hutchinson and his students. What they had in common was a willingness to use mathematical models while maintaining good standards of experimental design and field observation. This attitude led to a reappraisal of the logical steps of the earlier theories of population ecology, while retaining the broader appreciation of patterns which had supported them.[15] Hutchinson's emphasis on both logical deduction from facts and the search for larger patterns in nature was characteristic of his feeling, which one of his students felt he conveyed to them, that science is an artistic achievement.[16]

It was into this milieu that Robert MacArthur stepped in 1953. He had just received a master's degree in mathematics from Brown University and was interested in exploring some field of application. His father, John W. MacArthur, was a zoologist at the University of Toronto, so it would have seemed natural for him to enter some area of biology. He even had a coincidental connection to Raymond Pearl, for his father was one of the first to examine critically Pearl's duration-of-life studies back in the 1920s.[17] It was Slobodkin, however, who told MacArthur of the potential for using mathematics in ecology and who steered him in Hutchinson's direction.[18] Ecological studies were an ideal area for someone of MacArthur's intellectual leanings, because he combined a love of nature with a fascination for problems of resource allocation. Someone has recalled that even as a child he had been intrigued by the problem of how a cake might be shared among consumers of varying voracity.[19] This practical problem of household economy would become, when turned to the economy of nature, the problem of how resources were divided among competing species in a community. His academic studies were interrupted by several months spent working on mathematical problems for the U.S. Army, though he continued to think about ecological problems. When he returned in 1956 he began his thesis, which he completed in short order and headed off to Oxford for a postdoctoral year with David Lack in 1957–1958. Upon his return, he accepted a job in biology at the University of Pennsylvania, moving finally to Princeton University in 1965 as Professor of Biology.

As he left Hutchinson in 1957 for Oxford, the ideas of teacher and student had become closely meshed. The problems they addressed were traditional, but each was trying to ask questions in such a way as to suggest that answers could finally be found. Not that the answers would provide exact predictions of the kind that would bring nature under control. Such quixotic goals would be left to those who would soon be exploring the potential of the new computer technology, a technology that promised a more complex, realistic, and perhaps also predictive model of nature. Rather, their theories served an aesthetic function; they brought unity to ecology by showing how disparate facts could be understood in terms of general processes and concepts. The satisfaction of perceiving that unity was its own reward. After Cold Spring Harbor their styles began to diverge, as MacArthur's plan for the reconstruction of ecology began to take shape. Before long he had fashioned Hutchinson's basic design into an imposing structure which cast its shadow over almost every area of community ecology. What began as tinkering was turning into a dazzling piece of engineering.

But let us now go back to those first stages as they were revealed at Cold Spring Harbor in 1957. In response to the confusions evident at the meeting, Hutchinson decided that a more precise definition of the niche was desirable. A curious and otherwise ignored paper published in 1940 gave him the idea. Written by Edward Haskell, who had come out of the socioecological school of the University of Chicago, it drew attention to the need for better definitions of central ecological concepts.[20] Haskell envisioned a set of mathematical categories defining the environment, the organism, and the habitat, for which he coined an alarmingly large number of new terms: the habitat became a group of geometric "hyperbodies" with the organism as the "testbody," "either in actuality or in imagination." Hutchinson, though more restrained, took the central idea of this paper and, by applying a few principles of set theory, turned the niche into an "abstractly inhabited hypervolume of n-dimensions." A similar idea of a multidimensional niche space had been formulated by V. A. Kostitzin in 1935. Hutchinson knew of this work in the 1940s and later felt it must have also influenced him, though he did not remember the passage at the time.[21] His set-theory definition was not intended to be rigidly and eternally affixed to the concept of the niche; rather, it was a temporary device designed to sort out the muddled arguments surrounding Gause's principle. He explained its function:

> It is not necessary in any empirical science to keep an elaborate logicomathematical system always apparent, any more than it is necessary to keep a vacuum cleaner conspicuously in the middle of a room at all times. When a lot of irrelevant litter has accumulated the machine must be brought out, used, and then put away.[22]

The definition was used to make a distinction between two senses of the niche. The "fundamental niche" defined all the ecological properties of a species and represented the area occupied in the absence of competition. When competitors were present, species occupied a smaller portion called the "realized niche." (Later he generalized this distinction by referring to the two senses as "preinteractive" and "postinteractive" niches, respectively.)[23] Gause's principle stated that realized niches did not intersect. This was not a tautology, for there was no guarantee of its truth. By looking for places where it seemed not to apply, some interesting new facts could be found about how species did manage to live together. This was what MacArthur's doctoral research was about.

Hutchinson had suggested that a possible exception to the principle might be found in territorial birds, if the availability of territories effectively regulated population size to the level where competition between two species was unimportant. Each species therefore would inhibit its own population more than it inhibited its neighboring species. To test this hypothesis, MacArthur investigated a group of warbler species in New England.[24] He found that territoriality and differences in habitat and nesting date did appear to reduce competition. More surprising was his discovery that in the situation where the species would be expected to compete, that is, while they were feeding, there was a striking degree of niche specificity. Although all the species fed together in the forest, MacArthur found that different species tended to feed in different parts of the trees and had also developed slightly different feeding habits. To a remarkably fine degree they had divided resources between them. Gause's principle was upheld. The study also showed that fairly precise measurements of niche dimensions were possible. This research, a good demonstration in the Hutchinson mold of how a logical argument could be put to use, won the Mercer Award of the Ecological Society of America for the best ecology article of 1958. It was an impressive start to a promising career.

MacArthur believed that insights obtained from the field should be extended through mathematical analysis in order to generate new predictions and therefore new field research. In this respect he was no detached mathematician. He argued that a solid program of field research was ecology's strength; he was sensitive to the need to find predictions which could be properly tested; and he always derived his own insights directly from nature.[25] But in his enthusiasm to show the possibilities which he saw in a modern, reoriented ecology, he felt justified in sometimes publishing an idea in order to raise a problem, to show how interesting it was, rather than to solve it. He could also be most adamant on the limitations of an ecological style confined to narrow case studies without ever attending to a broader theoretical perspective: "Will the explanation of these facts degenerate into a tedious set of case histories, or is there some common pattern running through them all?"[26] His theoretical speculations did not imply a lack of respect for field work; on the contrary, he relied on his experiences in the field to teach him which questions to ask. But theories are generated much faster than experiments, so inevitably a large part of the field work and statistical analysis needed to test the reality of those theoretical patterns would be left to others. Several collaborators, both in the field and in purely mathematical work, filled in deficiencies in MacArthur's own background with their various areas of expertise. These collaborations helped set his ideas in a broader framework and contributed to their success. Above all, he worked to bring the complexity of nature to a level where it could be comprehended by the human mind. And in all such cases where simplicity is sought in order to make a start on a problem, it is also important to know when to distrust that simplicity.

The Broken Stick

How can community structure be analyzed? One way is to look at the number of species that a habitat can support, starting with an estimate of the relative abundance of each species in the community. Studies of relative abundance in ecology began in the early twentieth century.[27] Plant ecologists had long observed that most species in a community were relatively rare, while a few species were fairly common. A statistical study of species abundances provided one way to distinguish between communities in different geographical areas and in different stages of succession. The difficulty was to find a mathematical expression of the distribution of abundances which would hold for different communities. In animal ecology, early studies of abundance were aimed more at qualitative generalizations rather than statistical laws. Charles Elton had related an animal's size and abundance to its position in the food chain, an idea expressed in his familiar "pyramid of numbers."[28] Large predators were rare, while smaller animals at the bottom of the food chain were abundant.

As more data were collected, animal ecologists joined plant ecologists in searching for mathematical relationships in the abundances observed. These forays into statistical analysis marked a steady improvement in statistical knowledge among ecologists from the 1930s on. Largely responsible for this improvement was R. A. Fisher's text, *Statistical Methods for Research Workers*, which appeared in 1925 and was in its eleventh edition by 1950.[29] Fisher had himself contributed to the analysis of relative abundance in 1943, in response to some problems encountered by his zoological colleagues.[30] Their work stimulated Frank W. Preston (who was a glass technologist professionally) to develop an improved model in 1948, based on data from birds, which he hoped would predict abundances in communities that were inaccessible to study.[31] By the early 1950s the statistical analysis of abundances was beginning to attract wider interest among animal ecologists. MacArthur encountered it through an article published by Hutchinson in 1953, "The Concept of Pattern in Ecology," which ended with a short review of the problem and how it might be approached.[32]

MacArthur's method of analysis was a paradigm of his way of thinking

Robert Helmer MacArthur, 1930–1972
Photograph by Orren Jack Turner,
courtesy of Princeton University Archives

about all ecological problems. Previous studies had differed in their conclusions about what mathematical expression best fitted the data, but the general method was always the same. It followed what Ronald Ross had earlier called the a posteriori approach, a standard procedure when ample facts were available. The method consisted of fitting the data empirically to a statistical distribution which would, if it proved general enough, acquire the status of a law of species abundance. But the biological meaning of the equations so derived was uncertain. MacArthur set out to remedy this lack of connection between biology and model by turning the question around to argue logically from biological postulate to its mathematical expression.

The point of departure was once again Gause's principle of competitive exclusion and the current controversies over the role of competition in regulating populations. If species abundance was influenced by the presence of other organisms (through competition), then one would expect the community to be structured so that niches did not greatly overlap. If, on the other hand, abundance was controlled by some factor independent of the presence of other species (such as climate), then one should find overlapping niches in the community. The problem was to find a mathematical model to express these biological alternatives. MacArthur pulled a possible candidate out of the recent statistical literature, a model which was dubbed the "broken stick" distribution.[33]

According to his application of this model, the environment was compared to a stick broken simultaneously at randomly chosen intervals, with each segment representing the abundance of one of the species in the community.[34] This way of dividing the stick would correspond, he argued, to the situation of nonoverlapping niches. Alternatively, the abundance could be determined by measuring the distance between any pair of points thrown randomly onto the stick. In this case the species were independent of each other and niches overlapped. Each model predicted a different curve of abundance measured against the species rank in the community. By comparing these theoretical curves with data from the literature, MacArthur found that in some cases the closest fit was given by his first "broken stick" model. From this fit he concluded that species abundances were determined at least partly by competitive exclusion, a conclusion which was supported by his warbler studies. But the data did not fully confirm the model. Often the actual curves were too steep, with common species too abundant and rare species too rare. MacArthur did not reject his model. Instead he proposed that the cause of the discrepancy was the heterogeneity of the environment. If the community were subdivided into smaller homogeneous units, each separate unit did give a better fit. This meant that large communities were actually composed of two or more "broken sticks," each of different length.

Finally, a third hypothesis was meant to test whether niches were continuous or discrete units. Species were now compared with a number of urns into which particles (representing units of abundance) were tossed on independent random throws, each urn having equal probability. This hypothesis corresponded to the biological postulate of independent, discrete units. But the curve which was predicted did not fit the data well, suggesting that niches were more continuous than discrete.

The idea of connecting biology to mathematics in this way was immediately attractive, though there were several problems with this model. It was not certain that this randomly derived distribution would actually correspond to the biological situation MacArthur proposed. The model also raised too sharp a dichotomy between overlapping and nonoverlapping niches, and it only applied to equilibrium populations. Opportunistic species which fluctuated in response to the vagaries of climate were unpredictable. MacArthur considered them to be unimportant for understanding community structure for that reason. The creation and interpretation of the model reflected an early emphasis on equilibrium communities, where the importance of competition might reasonably be inferred. MacArthur was aware that it would be necessary to explore other types of dynamic processes in communities not in equilibrium. The distinction between "*r* and *K* selection," which I shall come to later, rests on his recognition that different questions must be asked of these different types of populations. But for the moment his emphasis leaned toward equilibrium communities; and even when recognizing that this was only part of the story, his preference for studying the equilibrium case was characteristic of his work as a whole.

MacArthur published his model in 1957 and again in 1960. By this time ecologists were also being educated about his ideas through Hutchinson's writings. At Cold Spring Harbor, Hutchinson had used the broken stick model to elaborate the analysis of Gause's hypothesis. A year later in his presidential address to the American Society of Naturalists, he again used the opportunity to weave together MacArthur's ideas and his own in a discourse designed to unravel the reasons why there were "so many kinds of animals" in a community.[35] The answer had been hinted at in his recent study of two species of water bugs found living together near a shrine to Santa Rosalia at Palermo, Italy. The "Homage to Santa Rosalia" that he delivered at the meeting ranged over a discussion of food webs, competition, the niche, and the character of the environmental mosaic. As in his concluding remarks of the year before, his tale hinged on the acceptance of the principle of competitive exclusion. Evidence for the view that niche boundaries were set by competition was, as he admitted, mostly indirect. But he felt that contrary views were due to misunderstandings and sloppy formulations of the problem. MacArthur's model provided a way to focus thinking about the problem of competition by uniting the analysis of abundance with an investigation of the structure of the niche. He fully realized the model was not general, but he believed that when it did work, for equilibrium communities in homogeneous environments, it described a real set of properties that were likely to be of biological interest.[36]

But it was not enough to notice that species occupied different niches and that these differences were related to small differences in structure and habit. This observation, earlier publicized by David Lack in his study of the Galápagos finches, invited the further question, now raised by Hutchinson, of exactly *how* different two similar species must be if they were to coexist. Hutchinson's answer was drawn from an empirical generalization, whereas MacArthur's relied more on formal mathematical analysis of the limiting similarity between two species.

The Evolving Niche

In 1956 W. L. Brown and E. O. Wilson had drawn attention to an interesting phenomenon, which they called "character displacement,"[37] observed among related species whose ranges overlapped. In the areas of overlap, the differences between the species were more marked than the differences where they did not overlap. Such character displacement seemed to be good evidence in favor of competition. Hutchinson wanted to know how great those differences were. He surveyed a group of coexisting species and found that they appeared to differ by a larger-to-smaller, linear size-ratio which varied from 1.1 to 1.4, with a mean of 1.3. He used this ratio as a rough empirical rule which, if it proved general enough, could possibly be used to indicate the level of competition among species.

MacArthur was also intrigued by the apparent constancy of this 1.3 ratio and began to investigate the problem of similar species along more theoretical lines. This question, or some version of it, occupied him for a good part of the next decade and more. The initial theory was explored in collaboration with Richard Levins, a geneticist who shared MacArthur's approach to modeling and with whom he worked extremely closely; in their collaborations it is difficult to separate the ideas of each one. They wanted to know what conditions set the limit to the diversity of coexisting species, or put another way, how closely species could be packed in an environment. The answer required looking not only at the character of the species, but at the environment as well. The success of a species depended on how efficiently it used the resources available to it, and this in turn depended on competition, on the size and distribution of the resources in relation to the animal, and on the stability of the environment.

The analysis which followed looked at two kinds of problems. First, given that niches were fixed, under what conditions could species coexist? Second, how would natural selection affect the position and breadth of the niche itself? From these questions they tried to predict the situations under which character divergence or character convergence would occur. Though the method clearly owed a great deal to Volterra and Gause, their evolutionary emphasis was an important advance. By relating the idea of fitness to features of the environment, such as stability, heterogeneity, or level of competition from other species, MacArthur and Levins were making an effort to join ecology and evolution by connecting the ideas of population ecology with those of population genetics. From a study of the niche dimensions in warblers to a general theory of the evolution of the niche, MacArthur was gradually expanding on Hutchinson's theme that evolution was played out in an ecological theater.[38]

Their refinement of the problem of coexistence provided a more subtle alternative to the bald dichotomy between overlapping and nonoverlapping niches of the broken-stick model. MacArthur had in fact already dropped this product of his youthful enthusiasm, for the most part, and had gone back to Preston's more general curve when discussing abundance.[39] He even renounced the broken stick in 1966, in response to some criticism of it, and expressed the hope that ecologists would cease to draw attention to it.[40] The renunciation did not last long. He continued to refer to the model in later articles, for it did have some value, even if it did not lead to the general theory of communities that MacArthur was after.[41] But even while the broken stick was being declared, however briefly, to be a bent reed, the underlying idea of setting out clear-cut alternatives was still firmly planted in MacArthur's mind.

Levins explained how the method used by him and MacArthur worked and how it was related to other modeling techniques.[42] Recognizing that no model could ever reflect the complexity of an ecological system, he identified three properties which a manageable model might hope to express: generality, realism, and precision. But no model could ever maximize all three. The models actually chosen for any situation were always compromises which sacrificed one of these features. Levins arbitrarily divided modeling strategies into three types, though not all approaches actually fitted so neatly into his categories. Models which sacrificed generality to gain realism and precision were used by economic biologists, who wanted testable predictions of specific trends and who could rely on computers to solve their unwieldy equations. Models which sacrificed realism to gain generality and precision were those such as the Lotka-Volterra equations, which could be used to get fairly precise results but were highly unrealistic in their assumptions. By looking at how nature differed from the models' predictions, the assumptions could be adjusted and realism increased.

The third strategy sacrificed precision to gain realism and generality. This was the method used by MacArthur and Levins: it was intended to increase the flexibility of their models, while avoiding a commitment to the unrealistic assumptions which had permeated Volterra's arguments. But the predictions could no longer be precise. They took the form of qualitative comparisons between types of organisms or types of environments: tropical versus temperate species, insular versus continental faunas, patchy versus uniform environments. These comparisons were expressed in a graphical form which they hoped would be easier to interpret than a set of equations. The method fitted very well with MacArthur's predilection for seeing the world in terms of dichotomies. The graphical method itself was actually an extension of Gause's graphical analyses of Volterra's competition equations. Discussed most fully in Gause's French monograph, these were well known to Hutchinson and had been recently included in Slobodkin's book of 1961. Though MacArthur referred to the method as a familiar one, it was not generally used in ecology because that part of Gause's work was still not widely read. The graphical method was brought into ecology chiefly by MacArthur and Levins's applications of it.

With any strategy there were problems of relating the model to reality and of knowing when the model could be trusted. As Levins explained: "There is always room for doubt as to whether a result depends on the essentials of a model or on the details of the simplifying assumptions."[43] To minimize this doubt, Levins advocated using several alternative models on the same problem, models which included different simplifications but which had a common biological assumption. If these different models reached a similar result, one could feel more confidence in that result: "Hence our truth is the intersection of independent lies." Levins dramatized this point by referring to such results as "robust." His article helped to popularize the idea of "robustness" in population biology, though the concept had already been in use for more than a decade. Because the use of this term has now become commonplace and is correspondingly vague in meaning, it is worth looking at its original meaning briefly.

The word "robust" was introduced in the early 1950s as a criterion of statistical testing.[44] One requirement of a statistical test is that it be sensitive to changes in the factors being examined; a test which met this criterion was said to be "powerful." But a test should also be insensitive to changes in any extraneous factors of a magnitude likely to occur in practice. A test which met this criterion of insensitivity was called "robust." The idea of robustness appeared in population ecology in a discussion of the exponential growth curve by J. G. Skellam in 1955.[45] His point was that such factors as the overlapping or nonoverlapping of generations, or short-term seasonal periodicity, might have no important effect on the overall exponential growth pattern of a population when measured from year to year. The growth curve was robust to the extent that it was insensitive to these factors.

In Levins's usage, it was not the models themselves which were robust (because they all carried incorrect assumptions), rather the consequences of the models were robust. William Wimsatt explains how Levins used robustness analysis as a way to discover empirical truths which are relatively free of the details of the specific models used:

> If a result is robust over a range of parameter values in a given model or over a variety of models making different assumptions, this gives us some independence of knowledge of the exact structure and parameter values of the systems under study: a prediction of this result will remain true under a variety of such conditions and parameter values. . . .
>
> Robust theorems can thus provide a more trustworthy basis for generalization of the model or theory, and also, through their independence of many exact details, a sounder basis for predictions from it.[46]

For Levins, a theory was made up of a cluster of models and their robust consequences. Levins was arguing here not only for the place of modeling in population biology as the only way to cope with complexity, but for the need to foster a pluralistic style of ecology to piece together a general theory of community structure from many sides. Of course he realized that the choice of different models or strategies would reflect conflicting goals and even conflicting aesthetic standards on the part of biologists. For this reason he regarded disagreements about methods as basically irreconcilable. But he saw the alternative approaches, even of opposing schools, as partaking of a larger "mixed strategy" which would fit different pieces of the puzzle of community structure into a coherent whole.

MacArthur's own views on how ecology should be done clearly reflected his aesthetic biases. Foremost was his insistence that the point of ecology was to discern the repeating patterns of nature and to interpret these in terms of general principles. Hutchinson held much the same views. But to see the patterns, MacArthur maintained, it was important not to be so closely enamored of any one part of nature to the extent that the sheer volume of facts obscured one's view. One had to step back to see the whole. The patterns were of interest because of their generality, yet as he admitted, "these kinds of general events are only seen by ecologists with rather blurred vision. The very sharp-sighted always find discrepancies and are able to say there is no generality, only a spectrum of special cases."[47] For MacArthur, knowing too much could be a disadvantage; it constrained the imagination. He conjectured that this had been one of the problems which had prevented the demographers at Cold Spring Harbor in 1957 from reaching any generalizations of interest to ecologists: "It may be very true that demographers know too much."[48]

The consequences of stepping back to catch the patterns was that precision was lost. Predictions were stated in terms of choices between alternatives, with several sets of alternatives operating at once. The choices were often between ideal types, comparable to the frictionless pulleys, conservative fields, ideal gases, and perfect crystals of physics, "but science is made up of such fictions."[49] Even if the ideal types were not found in nature, it was interesting to see how real species departed from the ideal.

The reference to blurred vision recalls another theoretician whose ideas greatly influenced MacArthur: R. A. Fisher. Fisher's *The Genetical Theory of Natural Selection* was reprinted in 1958 and soon set MacArthur to work deriving his own "fundamental theorem of natural selection" which would redefine fitness to allow for the effects of a changing environment on an organism.[50] The changes in this case were limited to changes in population density. In MacArthur's scheme, the carrying capacity of the environment (K in the logistic equation) replaced the maximum rate of increase (r) as the measure of fitness. The definition was unworkable, but the attempt reveals the enthusiasm with which the young MacArthur plunged in and tried to bring ecology and genetics together. Apart from the influence of specific ideas on MacArthur's thought, it was Fisher's approach to science, reflecting the mathematician's willingness to think in abstractions, that MacArthur also admired and sought to emulate.

Like Fisher, MacArthur's desire to formulate problems so as to generate more hypotheses had one other important consequence: it created a bias against historical explanations. Historical accounts took into consideration the vagaries of particular times and places. And although MacArthur was prepared to admit that sometimes history mattered, he was not comfortable with solutions that featured the unique rather than the general: "If the patterns were wholly fortuitous and due to accidents of history, their explanation would be a challenge to geologists but not to ecologists."[51] Purely historical problems would not, he believed, generate new hypotheses. For this reason he was always more interested in equilibrium populations, or as he put it: "It seems much harder to make interesting predictions about the more erratic populations."[52] Nor was he very keen on large descriptive undertakings which were meant to evoke the whole of nature, for that was merely to be swamped in detail. "Not all naturalists want to do science; many take refuge in nature's complexity as a justification to oppose any search for patterns. This book is addressed to those who do wish to do science."[53]

The problem was to choose a unit of manageable proportions, something in between a single species and an entire ecosystem. But just how to identify the right unit was not easy to explain. "We use our naturalist's judgment to pick groups large enough for history to have played a minimal role but small enough so that patterns remain clear."[54] MacArthur put great stress on the intuitive ability of the naturalist in choosing the interesting problems. It was partly for this reason that he distrusted the elaborate modeling done with the aid of the "new electronic devices." These might make calculations easier, but they gave little help in discriminating what problems were worth pursuing.[55] His intuitive sense of what was the best unit also made him impatient at the suggestion that terms such as "community" ought to be carefully defined before attempting to draw some general conclusions. "Irrespective of how other people use the term 'community'—and there are almost as many uses as there are ecologists—I use it here to mean any set of organisms currently living near each other and about which it is interesting to talk."[56]

Such definite opinions on what science was about could not help but create antagonisms. On the positive side, however, MacArthur's insistence that he had found the way to ask *interesting* questions also helped to create an image of a reinvigorated ecology on the verge of rapid expansion in new directions. One of the most successful of these new directions (though now the focus of controversy) was the theory of island biogeography which he developed in collaboration with Edward O. Wilson. Their theory is of interest not just as an illustration of MacArthur's method, but because it was the area where all of his interests converged and where the union of ecology and genetics seemed most promising at the time.

Island Biogeography

MacArthur was introduced to biogeography by Wilson, a zoologist at Harvard University, in the early 1960s, and it became his favorite part of ecology: "the part that combines the adventure of field work in varied places with the discipline of making nontrivial theory."[57] Beginning with the simpler environment of islands, MacArthur and Wilson advanced a new theory of biogeography in 1963.[58] Their theory was made possible by the faunistic data collected in the 1930s and 1940s by K. W. Dammerman, Ernst Mayr, Bernhard Rensch, and E. Stresemann. A fuller treatment of the subject appeared in 1967 in *The Theory of Island Biogeography*.[59] This book also served as a manifesto for a reform which, they hoped, would "galvanize biogeography and have extensive repercussions in ecology and evolutionary theory."[60] Its purpose was to nudge biogeography out of its "natural history phase" with its descriptive histories of individual taxa and biota, toward an experimental and theoretical phase, where new hypotheses deduced from theoretical arguments would be tested in the field. The first principles came from population ecology and genetics. The theory was admittedly crude and it sometimes relied on concepts invented as they went along, but its function was to generate hypotheses, not final answers. And that meant turning away from history, as least for a time. They argued that one of the reasons biogeography had not been made theoretical was that previous research, being mainly taxonomic, had been dominated by a historical viewpoint: "The conventional issues relate to special places and special groups of plants and animals. . . . The historical solutions have tended to be satisfying in themselves and have not encouraged generalizations."[61]

Their theory was based on the observation that there was an orderly relationship between island area and species diversity. One way to explain this relationship was to correlate diversity with various features of the environment: degree of isolation, area, elevation, complexity, and so on. But useful as such studies were, MacArthur and Wilson found this descriptive approach not suitable for creating the kind of hypotheses they were after.

Their alternative was to interpret diversity (meaning the number of species only) as a dynamic equilibrium between immigration and extinction. Ernst Mayr had actually arrived at a similar view using different reasoning in the 1940s, when he concluded that the turnover of fauna on islands was much faster than had been previously thought, as a result of frequent extinctions of these vulnerable populations.[62] Later he commented that his idea was "long ignored" until it was "transformed into graphs and mathematical formulae" by MacArthur and Wilson.[63] The equilibrium view of island diversity had also been anticipated by Frank Preston in 1962.[64] MacArthur and Wilson's theory agreed in part with Preston's but contained some predictions he had missed. Their use of the equilibrium concept was different from Mayr's in that the whole point of their model was to make the vagaries of local history seem less important for understanding diversity. Of course history was crucial if one wanted to know the exact taxonomic composition of the species, but MacArthur and Wilson were at this stage not interested in the identity of the species, only in some general observations about why numbers varied in different places.

Their model was expressed graphically by two curves: a decreasing curve of immigration rate and an increasing curve of extinction rate, measured against the number of species present on the hypothetical island (Figure 8.1). The point of intersection of the curves gave the equilibrium number of species. This model was comparable to interpreting the number of people in different cities, for instance, not in terms of the history of a city's growth, but simply as a balance between the birthrate and immigration and the deathrate and emigration of the population. The islands were similarly considered to contain "populations" of species, with diversity seen as the result of a "birth and death process" of species. But the model had several imperfections. The exact shape of these curves was unknown, nor did they know whether the same curves would hold for different islands and different times. Quantitative predictions of the number of species were therefore not possible. Finally, the criteria for distinguishing valid cases of immigration and extinction were not set out. Though they were aware of these problems, they felt that the model's value was in its qualitative predictions, some of which were not obvious on commonsense grounds alone. Verification of such predictions would be an important test of the validity of the equilibrium model. The data needed to provide thorough tests were not yet available, although for the time being they were able to muster enough data for provisional support. MacArthur and Wilson forged ahead. The rest of the book was devoted to working out the ramifications of their model for the ecology and evolution of colonizing species.

First to be considered were the properties which made a good colonizer. Their analysis involved relating the life history of a species to its chances for success in a new area, for a species had to have the right life-history strategy if it were to survive. Earlier work by Lotka had given the formulae for calculating the intrinsic rate of increase (r). More recently, his methods had been brought into ecology and genetics by Lamont Cole and Richard Lewontin.[65] From a consideration of the life-history parameters of the population—rate of increase, birthrate, deathrate, and Fisher's concept of reproductive value—MacArthur and Wilson were able to draw a hypothetical portrait of the "superior" colonist. They next considered the converse situation of why potential colonists were sometimes prevented from invading a new territory, no matter how many times they tried. This problem was simply a version of the question Hutchinson had posed back in 1959: Why were there so many species in a community, and not more?

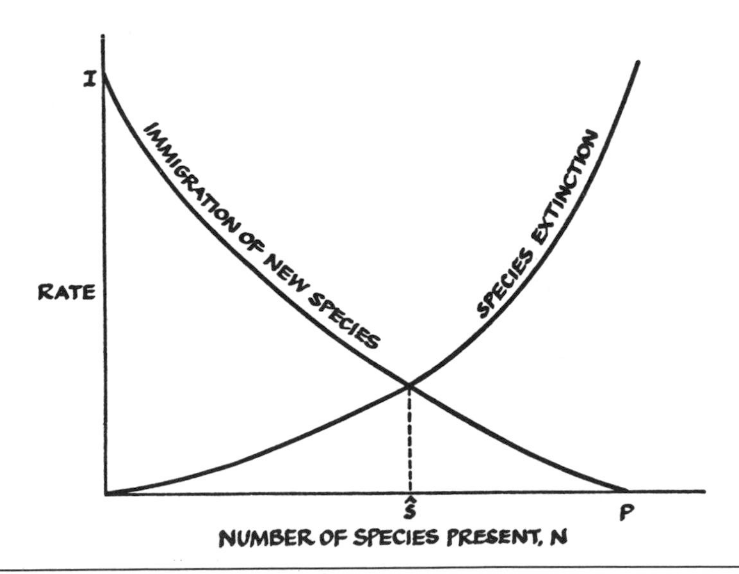

FIGURE 8.1. Graphical representation of the equilibrium model for a single-island biota. The equilibrium number of species, $\hat{S}$, is at the intersection of the hypothetical immigration and extinction curves. The slope of these curves was assumed to vary according to size of the island and its distance from the source of immigration. (From R. H. MacArthur and E. O. Wilson, *The Theory of Island Biogeography*, p. 21, © 1967 by Princeton University Press. Reprinted by permission of Princeton University Press.)

MacArthur had been absorbed in the formal analysis of this question ever since then. He and Wilson built on Volterra's observation that there could be no more predator species than prey species in a community, to which they added a refinement based on the idea of "grain" in the environment.

Originally developed by MacArthur and Levins, the concept of grain was a way of relating resources to the size and the search strategy of the predator. If resources were mixed in such a way that the predator found them in the proportion in which they occurred in the environment, then they were said to be "fine-grained." If, on the other hand, the predator could find one resource in more than its natural proportion, the resource was "coarse-grained."[66] The purpose of their analysis was to show that diversity depended not only on the numbers of resources and predators, but also on the "graininess" (relative size and distribution) of resources. Diversity also depended on other factors, such as the level of competition present, predation pressure and disease, and the minimum number of individuals needed just to maintain a population.

Their analysis led them to look closer at the dynamics of colonization, that is, the short-term ecological changes a species would undergo as it moved from one environment into another. The changes were of three types: ecological compression, release (expansion), or shift. Compression and expansion could be readily observed, the main problem being to figure out just what aspects of the niche (food or habitat size, for instance) actually underwent the initial change. Ecological shift was harder to prove, but MacArthur and Wilson theorized that species should place themselves so as to minimize competition, although it was not clear how such shifts could be measured in relation to competition pressures. The point, however, was to indicate that there were many ways, some of them unexpected, that species could be excluded from islands.

A study of the function of small "stepping-stone" islands in promoting colonization led to a final discussion of evolutionary changes on islands. Again the emphasis was on making hypotheses, starting with a definition of fitness and then suggesting how different life-history strategies would act to maximize fitness. But a major modification of the usual idea of fitness was first required. Fisher, Haldane, and Wright had defined fitness in terms of the rate of increase, but these measures assumed an unchanging environment. MacArthur had long been dissatisfied with this idea of fitness and as early as 1962 had proposed an alternative which took into account environmental changes.[67] But the only *predictable* environmental change was that caused by crowding as the population grew, assuming the population grew logistically. MacArthur's new definition included only that aspect of the environment. To analyse fitness, he merely transferred Gause's analysis of competition between two species to the level of the genome. Now, however, the competition was between two alleles, not two individuals, and the *K*-values represented the carrying capacity for different sets of alleles. The outcome of this evolutionary competition between alleles paralleled the various outcomes of Gause's analysis of competition (Figure 8.2). The concept of fitness was therefore analogous to the process of competition in a crowded environment.

MacArthur had spent considerable time developing his alternative concept of fitness, but in making the idea of fitness more realistic he had also made it unmeasurable. Unlike the old definitions, MacArthur's fitness could not be independently calculated for each species, based on the knowledge of its life history, for it was not a biological characteristic of the species as such, but a characteristic of the species in relation to a given environment. The measure would change according to circumstances, and its meaning was slippery. However, MacArthur and Wilson were not after quantitative predictions. They used the new idea to make a qualitative distinction between selection in a crowded environment and selection in an uncrowded one. With no crowding, populations capable of growing the fastest would be favored, and evolution would encourage productivity; they called this process "*r* selection" But in a crowded environment, evolution should favor more efficient conversion of food into offspring, or "*K* selection." In the context of biogeography, newly colonizing species should have *r*-selection, whereas an established species would shift to *K*-selection. Therefore comparisons of similar species in different environments should reveal the development of different strategies favoring productivity or efficiency. As with the other models in the book, these predictions were in the form of qualitative comparisons between ideal types. The book ended with a discussion of the possible course of evolution as seen by character displacement or its converse, character release, when species moved into new environments.

This brief sketch indicates some of the strengths and weaknesses of the book. The connections between the mathematical arguments and the biological hypotheses they suggested were sometimes tenuous. The biological conclusions were not always wholly dependent on theory. MacArthur and Wilson noted several places where their conclusions had been anticipated by others on the basis of different lines of reasoning. But the fact that their hypotheses had been independently derived tended to lend credence to their theories. Other predictions were surprising and counterintuitive, but their theory's success on the reasonable hypotheses suggested that perhaps the surprising ones had merit as well. They admitted that their ideas were expressed in terms of new concepts invented as they went along, but whose meanings were not clear. Such shorthand expressions as "*r*- and *K*-selection," while clearly connected to meaningful comparisons between evolutionary strategies, nevertheless were especially susceptible to misunderstanding. Finally, even the central equilibrium hypothesis had

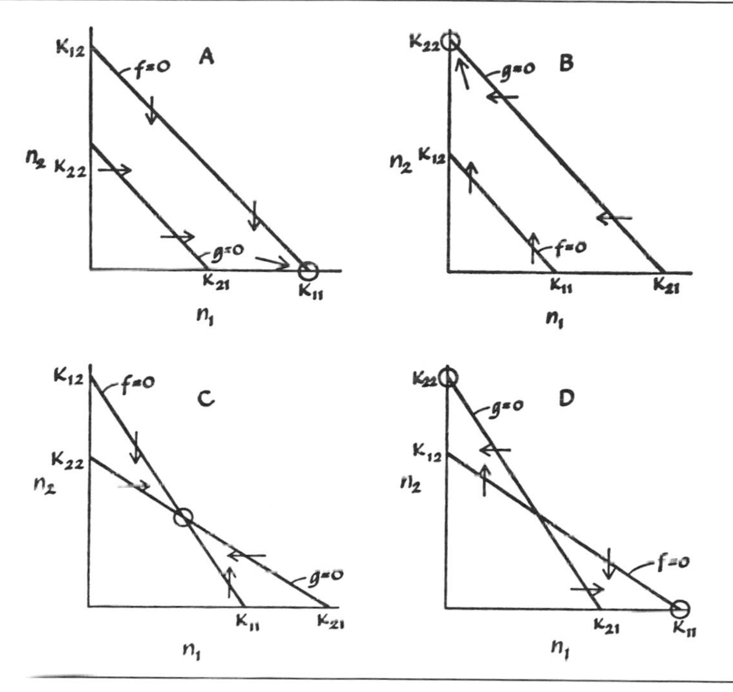

FIGURE 8.2. The principle of competitive exclusion brought to the level of the genome. MacArthur and Wilson's depiction of the four possible outcomes of density-dependent selection of two alleles. Their treatment is identical to Gause's (1935) analysis of competition using Volterra's equations, later described in Slobodkin's (1961) book. (*a*) Allele 1 will always oust allele 2; (*b*) allele 2 will oust allele 1; (*c*) both alleles will coexist indefinitely; (*d*) whichever allele is present in greater number will oust the other. (From R. H. MacArthur and E. O. Wilson, *The Theory of Island Biogeography*, p. 147, © 1967 by Princeton University Press. Reprinted by permission of Princeton University Press.)

not been tested to see how general it was. Mayr had pointed out in 1965 that birds, for instance, might be a unique case of the rapid turnover of species, for in some areas plants and insects associated with plants showed a different situation.[68]

But what MacArthur and Wilson's book was advocating was a new style of ecology, and it was the success of the style rather than any single hypothesis which would determine its merit. The style was based on a dialectical relationship between formal theory and experiment. The theory would provide a reference point, a series of idealized "what if?" scenarios, which would be a standard by which to judge deviations in the real world. The hypotheses would consist of comparisons between different types of species and different places. Experiments would test the hypotheses and, if necessary, help to reformulate the theories. The book was brimming with hypotheses; there was plenty of work yet to be done. It ended with a call for detailed studies of selected species to refine the arguments. Ecological and evolutionary processes would be integrated as far as possible. And the patterns of the whole—the patterns of species diversity—would be explained in terms of causes regularly operating within the community, such as competition.

As a stimulus to research, their book was a success. Serious attacks on its principles would be a decade in coming. Even those who disagreed with the mathematical approach found the originality of the analysis refreshing. David Lack, for instance, was not comfortable with what he called MacArthur's "distant" view, which concentrated rather abstractly on the number of species instead of particular species of a place.[69] He preferred what he called a "close" view, which drew conclusions from "intensive study of a particular situation in all its aspects." Nevertheless, he had reread MacArthur and Wilson's book "more times than any other on biology, because I keep finding further new and revealing biological insights in it."[70]

Lack's last book, *Island Biology*, was an attempt to develop his own viewpoint and to clarify the relation of his ideas to MacArthur's. Written in haste during a terminal cancer, it was completed only after MacArthur's death, also of cancer, in 1972 at the age of forty-two. MacArthur had also hastened to write down his ideas before he died. These appeared in *Geographical Ecology*, a book which encapsulated all of his ideas about science. In a final short chapter on the role of history, he unrepentently forged an alliance between ecology and physics; the dichotomy he set up there captures very well how he thought about ecology:

> The ecologist and the physical scientist tend to be machinery oriented, whereas the paleontologist and most biogeographers tend to be history oriented. They tend to notice different things about nature. The historian often pays special attention to *differences* between phenomena, because they may shed light on the history. He may ask why the New World tropics have toucans and hummingbirds and parts of the Old World have hornbills and sunbirds. The machinery person may instead wonder why hummingbirds and sunbirds, despite their different ancestries, are so similar. He tends to see *similarities* among phenomena, because they reveal regularities.[71]

Then MacArthur ended with a brief bow to the places "where history leaves an indelible mark even upon the equilibria so dear to the ecologist." Never content to merely put forth new hypotheses, MacArthur was also trying to redefine ecology as a science. Whether others would accept this redefinition would be a matter for future debate, but for the moment his ideas would leave an immediate mark on the character of American ecology, causing one reviewer to enthuse over the "provocative brilliance" which had led "to his being the most cited ecologist of his time."[72]

Consolidation and Defense

MacArthur's rapid rise to fame, followed shortly by notoriety, owed much to the ability of his ideas to generate new research. His success partly stemmed from the fact that he did derive his ideas from field experience; from his warbler studies, from later work on foliage diversity in bird habitats, from field trips to Central America to investigate problems in biogeography. Much of this work was done in collaboration with students, colleagues, and members of his family. It did not always show up in his publications. Often he would develop a theoretical argument without revealing the biological observations which had stimulated it. But his speculations were based on biological insights which others could appreciate. As Hutchinson said, "Robert MacArthur really knew his warblers."[73] And he did care about making testable predictions. Indeed, he has been credited with single-handedly turning ecology into a hypothetico-deductive science, as though ecologists would not have learned to test hypotheses otherwise.[74] This claim is exaggerated, but it expresses a more general belief that ecology in the 1960s was being remade and that MacArthur had a large hand in the remaking. George Salt recalled that, within the American community as a whole, "research became focused rather than diffuse and idiosyncratic. The entire intellectual climate in ecology and evolutionary biology became more taut and incisive. . . . A unified theory of evolution based on competition and adaptation seemed to be possible."[75]

MacArthur's impact may be roughly gauged from the proliferation of articles in the *American Naturalist* dealing with niche breadth, species packing, competition, biogeography, and foraging strategy among predators. *Geographical Ecology* was full of citations of new field work inspired by his methods, some of which had helped to rekindle MacArthur's briefly waning interest in geographical ecology by showing him how many ideas could be tested. Several new textbooks based on MacArthur's principles helped to disseminate his ideas.[76] His work also inspired further theoretical studies in ecology.[77] Wilson explored some applications of the theory of biogeography to the design of natural preserves, though he was criticized for too hasty application of what was still an unproven theory.[78] This was one of the few attempts at applied ecology. As Levins noted, theoretical ecologists located mostly in universities tended to see applied problems as "uninteresting."[79]

MacArthur's success also owed a great deal to his shrewd awareness that theoreticians had to successfully colonize a piece of the ecological territory. Writing textbooks was one good way to establish oneself, at least as far as the next generation was concerned. In 1966 MacArthur co-authored a text with Joseph Connell; his final book on geographical ecology was also intended as a textbook.[80] To provide an outlet for theory, he helped to establish the journal *Theoretical Population Biology* in 1970. Three years earlier he had started a monograph series devoted exclusively to theoretical ecology, his book with Wilson being the first in the series, which continued at a rate of about one book a year. His several collaborations and the generosity with which he shared his ideas created excitement about the possibilities of theoretical ecology among his colleagues. At least two of his collaborators, Hutchinson and Levins, actively promoted his methods for a wider audience. MacArthur's personal style—his quiet intensity and imaginative curiosity, his logical way of thinking through a problem—soon created a circle of loyal admirers.

MacArthur's enthusiasm carried over to the printed page as well, as he set out repeatedly the appealing image of the new, vigorous ecology making a sharp break with the stodgy past. "One characteristic of the history of ecology is that students tackle completely new and unexpected problems rather than tidy up those left by their teachers," he wrote.[81] This was an unexpected assertion from someone whose own work grew out of his teacher's, and who was so conscious of the role of textbooks in shaping students' views. But he knew the attractions offered by a discipline where even a beginner could create a stir by doing something new: "It is one of the enticing aspects of community ecology that it is so young and rapidly changing and that most investigators can make a substantial contribution."[82] In this sense also he eclipsed history. Each time he declared that his ideas would soon be obsolete (as he often did), he implicitly invited someone to step in and take his place. He fully admitted that many of his theories had serious shortcomings: they neglected individual differences, were often valid only near equilibrium, and ignored temporal changes in the environment. But MacArthur had no patience with those who doubted that a "satisfying theory of ecology" could be created simply because the world was too complex. "These are largely self doubts."[83]

MacArthur was by no means alone in drawing sharp boundaries between the descriptive ecology of the past and the mathematical ecology of the future. The times were full of great expectations of ecology's potential. People were more willing in the 1960s to look seriously at a variety of mathematical approaches: some fairly direct extensions of past work; others entirely new to ecology. In genetics Richard Lewontin, echoing MacArthur, pushed for the creation of a new, rigorous, and predictive science which would brush aside the "stamp collectors" whose influence had even "tended to degrade population biology as a science."[84] In population dynamics, which was concerned with short-term changes in economically important species, it was realized that mathematical modeling would be an important part of the field's future. Even so outspoken a critic as W. R. Thompson had softened his views by the 1960s and returned to

Volterra's work to see how theory might better be integrated with practice.[85] In a review of mathematics in ecology, Kenneth Watt drew attention to the role computers would soon have, as ecologists turned to detailed simulations of ecological interactions.[86] Computers and new technology would, he hoped, free ecologists from the drudgery of data collection and analysis, giving them more time for "high level thought."[87] He pointed out that ecology could well make use of the various types of mathematics designed to cope with complex systems in areas such as economics, transportation, the military, and engineering.[88] Game theory provided another source of models and methods.[89] The growth of demographic analysis and the emphasis on energetic relations encouraged "strategic" thinking, where the organism was thought to be designed according to some principle of optimization. Information theory had also sparked interest among ecologists, for it promised a unifying principle as general as the second law of thermodynamics.[90] People pointed to analogies between entropy in physics, information in communications theory, and diversity in ecological communities, and hinted broadly at some deeper significance to these analogies. The analogies were not always valid. Entropy and information (in the technical sense used in information theory) both measure the degree of disorder or randomness in a system, whereas ecological diversity is not always understood in quite the same sense. MacArthur used the analogy to suggest that in more diverse communities, material would flow through the food web along a greater number of alternative pathways. To the extent that this suggests greater randomness, the analogy is more suitable.[91] Not all these mathematical strategies were suited to the same goals; each possessed virtues the others lacked and shortcomings the others remedied. But there was enough interest in exploring new methods to ensure that MacArthur started with a receptive audience.

While MacArthur had lived to see his ideas embraced by a new generation of ecologists, he had also felt the pressure of mounting criticism. His qualitative predictions were too general; they appeared to gloss over the detailed insights into nature which intrigued the ecologist. MacArthur was interested in those details; he just believed that they had to be interpreted in the light of theory. Others were wary of the speed with which these theories were being incorporated into ecology. A text by one of his students was criticized for threatening to "transmute today's speculation into tomorrow's dogma."[92] In the second edition of this text, the author excused himself in the face of this accusation, by claiming that a text necessarily had to "gloss over some controversial areas and untested ideas,"[93] otherwise it would confuse its student readers.

MacArthur himself had warned against premature generalizations. He knew that present theory was incomplete, but he also believed it would be only a matter of time before final answers in ecology would be at hand. His prediction of the future of ecology imagined a world fully ordered according to general principles in one vast classification scheme:

> I predict there will be erected a two- or three-way classification of organisms and their geometrical and temporal environments, this classification consuming most of the creative energy of ecologists. The future principles of the ecology of coexistence will then be of the form "for organisms of type A, in environments of structure B, such and such relations will hold." This is only a change in emphasis from present ecology.[94]

Though MacArthur had always stressed the provisional nature of his hypotheses, he did have faith that his overall point of view would prevail. His prediction of the future sounds very much like a Kuhnian "mopping up exercise" following a MacArthurian "revolution."

At times his readers seemed too hasty in wanting to see this prediction fulfilled. They began to classify nature before the meaning of the theory was established or adequately tested. Pianka's text included a "periodic table of niches," which tried to order the whole range of organic nature according to trophic level and place on an "r-k selection continuum."[95] The fact that this table had already been criticized as meaningless did not stop the author from including it in the second edition of the book.

Theoretical ecology seemed to be forming a new "establishment" a bit too quickly, and ecologists were made skeptical and uncomfortable in the face of its claims. Even people with good mathematical training were finding some of the new theoretical arguments hard to fathom.[96] George Salt described the sharp polarization that formed anew in the late sixties: "As the mathematics used became more esoteric, the field biologists found themselves either unwilling or unable to keep abreast while at the same time maintaining their necessary tools in systematics, statistics, physiology, and morphology. . . . The fraternity became splintered into factions with lofty or derogatory titles depending on who applied them. 'Empiricists' and 'theoreticians' were the most polite."[97] An editorial in *Ecology* in 1974 suggested that cliques among some theoretical ecologists were resulting in discrimination against empirical studies not falling within the approved theoretical boundaries.[98] The theoretical group felt equally besieged and impelled to defend the necessity of theory. These debates had a positive benefit in that they encouraged a more careful look at the function of models in ecology and the relationships between the different models.[99] Their negative effect was to promote an increasingly defensive response on both sides. The battle lines were already well drawn by 1972 when MacArthur, surveying the state of ecology, lamented the rifts which had developed between schools:

> Unfortunately there are propaganda efforts by insecure members of the various schools aimed at others, and it would not be the first time in history if

> one of these efforts succeeded in temporarily putting one school out of favor. . . . In the interests of freedom and diversity, even these destructive attacks must be tolerated, but it is well to recognize that they tell us more about the attacker than the attacked. However, it is a pity that several promising young ecologists have been wasting their lives in philosophical nonsense about there being only one way—their own way, of course—to do science.[100]

Two years later, the writers of the *Ecology* editorial felt the need to remind theoretical ecologists of MacArthur's words. The argument for diversity and coexistence was certainly the reasonable position to adopt. Nevertheless, some of the conflict had arisen because MacArthur had never hesitated to give his opinions about what kind of science he thought was worthwhile. His critics were oversensitive because they perceived him and his disciples to be advocating their own style as the only scientific one; to MacArthur's rhetoric they responded in equal kind. The definite biases of MacArthur's approach could only aggravate the differences between those, like himself, who had their imaginations fixed on the large patterns, and those who preferred to scrutinize the details of each case. The dying MacArthur may have adopted the role of peacemaker, but he was also the reason the war persisted.

His bold and novel ideas captured people's imaginations because they promised a new enlightened age of ecological science. His influence was powerful because he was not afraid to be biased, to focus on problems which seemed tractable, and to leave the others aside. The most important bias in his science came from a dichotomy formed early on, which was itself a reaction to the controversies on population regulation fought out at Cold Spring Harbor in 1957. The choice was simple: either organisms interacted, with the levels and degrees of interaction being responsible for community structure; or they were independent of each other, in which case external factors such as climate settled the community's appearance. The second alternative relied heavily on fortuitous events and historical explanations. MacArthur preferred to explore the possibilities of the first alternative, without recourse to history if at all possible. His research helped to crystallize thinking about a certain class of problems involving competition and resource allocation; it generated a wealth of new hypotheses. To some it may have appeared that the mathematical arguments established the truth of certain observations, much as Volterra's arguments appeared to do a generation earlier.[101] And just as that earlier generation had discovered, closer scrutiny would necessarily force some revisions in the theories.

Nelson Hairston used some ideas from information theory to attack the logic behind the broken-stick model in 1959.[102] He still shared MacArthur's belief that competition explained certain differences in the distribution and ecology of similar species, however. By 1964 he had come to reject the model because it did not fit the data well, just as MacArthur was also coming to conclude that the model had outlived its usefulness.[103] But once embedded in the literature, ideas do not disappear at a blow. Hairston had to return to the attack in 1969, this time more forcefully, when he found that the model persisted in ecology despite its inventor's wish to the contrary.[104] Now he argued that no biological significance could be found in the fact that a set of data did not fit the model. But more recently, in a survey of mathematical models of abundance and diversity, Robert May showed that the broken-stick distribution would be expected when there was an even division of some major environmental resource among the community's species.[105] So while the distribution does not validate MacArthur's original argument, it does have some biological significance when it occurs.

Other criticisms focused on the interpretation of competition, so central to MacArthur's science. Richard S. Miller, formerly a student of Charles Elton in the early 1950s, published a thoughtful review and critique of the meaning of competition in 1967.[106] Competition was a notoriously ambiguous concept. Nobody doubted that it did occur, but the problem was to know where, and when, and with what degree of strength. MacArthur had been fully aware of the need to distinguish between strong and weak competitive interactions in the field,[107] but so much of his theory was based on the assumption of competition that experimental evidence from the laboratory in support of his hypotheses was often taken to confirm the prevalence of competition as well.[108] By the late 1970s the controversies over competition had flared to considerable intensity, and they continue to arouse passions to this day.[109] Part of the problem was that statements introduced as tentative generalizations, meant to stimulate more research, became fixed assertions when repeated later on. Louise Roth has uncovered how Hutchinson's 1.3 ratio became transmuted by others into a biological constant over its twenty-year history.[110] The scientific criticisms take basically two forms: one, a call for more careful field experiments to test basic assumptions, such as the strength of competition; the other, a call for more careful statistical analysis of data on species distributions.

A long-standing source of conflict arises from the difference between arguments which infer the importance of competition indirectly and those which seek to test its strength. Experimentalists sometimes feel their work is not appreciated by those who produce theories. For instance, Hairston has recently and reasonably argued the experimentalist's position that inferences drawn from patterns to underlying processes need more rigorous experimental assessment.[111] But he went on to suggest that Hutchinson had implicitly argued that an experimental method should not be needed to analyze competition. Such an implication need not be drawn from Hutchinson's writings; however the comment points to an important problem in ecology: the fact that field studies cannot keep pace with theories generated at a much faster pace.

A different type of criticism focuses on the methods of testing hypotheses. Daniel Simberloff, a former student of Wilson's and an advocate of the equilibrium theory of island biogeography, later became highly critical of this same theory.[112] His criticisms are based on a principle which students nowadays encounter in their elementary statistical courses, the creation of null hypotheses. The idea is to propose an alternative hypothesis to the one being tested and to set up an experiment which would try to falsify the alternative. In this case the alternative hypothesis is to assume that perceived patterns are entirely random. Only if the association can be shown to depart from randomness, it is argued, is it permissible to use explanations involving competition or other interactions. Falsification of a null hypothesis is seen to be a more objective test than directly trying to confirm the hypothesis in question. But the method of falsification is not free of error. Several people have pointed out in reply to Simberloff that null hypotheses are based on null models and these also use biological assumptions.[113] Therefore one must use due caution even in creating null models. Another problem is that by falling back continually on a position of scientific agnosticism, one may never create anything. This aspect of the debate recapitulates at heart the Cold Spring Harbor Symposium in 1957. Simberloff sees himself as part of a modern "probabilistic, materialistic revolution" in ecology, but this conflict is revolutionary only in the sense of traveling in a circle.[114]

The response to Hutchinson and MacArthur continues and provides adequate testimony to their pervasive influence on modern ecology. Though it is too early to see just where current debates will lead, it is evident that periodic stocktakings of the role of mathematical models in ecology continue, as they have in the past, to influence standards of hypothesis making and testing. Implicit in these discussions of methods and principles is the question: How many schools of ecology can peacefully coexist? But they also serve as reminders that nothing can replace an intimate knowledge of some part of the real world; that the naturalist is on equal terms with the mathematician. Robert MacArthur, even with the biases inherent in his view of science, would not have disagreed.

Conclusion

From 1920 to 1970 theoretical ecology moved from the margin to a permanent and more central place in ecology. The range of problems covered by theoretical analysis can be seen by a glance at some of the topics in the Princeton monograph series started by MacArthur: evolution in changing environments; adaptive geometry of trees; stability and complexity in model ecosystems; group selection in predator-prey communities; food webs and niche space.[1] Problems in resource management continue to rely on mathematical modeling techniques.[2] Plant ecology was revitalized in the 1960s by the transfer of population concepts taken from animal ecology and applied to the study of plants.[3] Sociobiology, launched by E. O. Wilson in 1975 amid much controversy, was an attempt to synthesize population ecology and population genetics, with the addition of an important ingredient missing from population theory, the study of behavior.[4] Epidemiology has again become the focus of a great deal of new theoretical work.[5] Mathematical ecology has even been applied to the analysis of turbulence phenomena in physics, "a pleasing inversion of the usual order of things," as Robert May remarks.[6] Much attention is still paid to the unrealistic assumptions behind the models, but there have been steady gains in the sophistication of the models; in the appreciation of how different models stand in relation to each other; in the understanding of the properties of the models; and in the idea of using clusters of models to attack a problem from different sides. This understanding has largely been the product of greater collaboration between mathematician and biologist. Mathematics has helped to uncover a vastly richer ecological world.

The rapid recent growth of mathematical ecology was preceded by four decades of sporadic efforts to establish a mathematical science of populations. The successes and failures of these years determined what ideas would be handed down to form the core of the modern science. The history consists of a few hits and a great many misses. The majority of the ideas published in the early years lay ignored or forgotten. Some were later rediscovered. Mathematical ecology has rightly been described as, to some degree, a process of reinventing the wheel. Other successes depended not so much on the intrinsic merits of an idea, but on how it was promoted; on how accessible it was mathematically; on how it was disseminated through personal connections. When mathematical ecology later achieved respectability, its content was largely the result of accident; nor should this fortuitous and uneven development surprise us.

Underlying the fortuitous sequence of events there are repeated patterns. The history of population ecology is one of a continuing dialectic between mathematician and biologist. The mathematician, trained in the physical sciences, sees equilibrium and uniformity, often at the cost of ignoring individual differences. The biologist, on the other hand, may see individuality; heterogeneity; constant, unpredictable change. The crucial difference between these two points of view is that the first is ahistorical, the second historical. But the processes one wants to understand—succession, geographical distribution, evolution—are historical.

It is possible, having begun with a simple mathematical model, to start putting history back into it in the form of more realistic assumptions, such as time lags. This is the route followed by economic biologists, but the result is very soon mathematically intractable. As Richard Levins pointed out, the more precise a model becomes, the less general it is. But often the purpose of using theoretical models is to arrive at general principles. The mathematical ecologist argues that this method is necessary to the pursuit of science. The biologist counters by pointing out that sometimes history does matter. The difficulty of combining these two approaches to nature, of realizing when historical answers are the correct ones, means that this dialogue between mathematician and biologist may continue much as it has in the past, although collaborative efforts will provide some resolution.

Mathematical modeling is not the only source of an ahistorical (or equilibrium) view. It would be interesting to know whether such a division between points of view can be found in other, unexpected areas of biology; whether there are parallel divisions in intellectual traditions outside of science; and how these may reflect changes in society at large. In an analysis of British geology in the nineteenth century, for instance, Louis Rosenblatt has drawn a parallel between geological theories, which he sees as representing both historical and equilibrium styles of reasoning, and interpretations of human history around the same time.[7] Certainly the ahistorical view was common in natural history long before the creation of ecology. Before Darwin, unity and harmony were seen as being imposed upon the world by a Creator whose power, wisdom, and goodness could be read directly from nature. This was the tradition of William Paley and later of the Bridgewater Treatises. And though Darwin is said to have removed the argument from design from natural history, the ecologists who turned natural history into science soon found other ways to restore the balance in nature.

The need to impose harmony on nature accounts for the appeal that

Herbert Spencer had for ecologists, and explains why organismic metaphors were taken seriously. Nature was still governed by laws, but these were interpreted in terms of a variety of metaphors which replaced the divine creator: physical and chemical analogies, machine metaphors, and finally metaphors drawn from human society, from game theory, engineering, and economics. In nature we see ourselves. All these constructs served to impose upon nature the regularity and structure needed to bring order out of chaos. Empirical observations suggested that these models, for all their simplifications, had at least some basis in reality.

Apart from the direct transfer of ideas from one branch of science into another, in the form of analogies, there is another source of ordering or unifying concepts which deserves mention: teaching. The act of having to order facts for the classroom can sometimes result in a rigid, formalized view of a subject. Many people have drawn attention to this process; once having arranged a topic in order to be able to teach it, it is easy to take the next step and begin to believe that the structure one imposes really exists. It is difficult to assess whether this problem really enters into the histories I have discussed, but Charles Elton believed back in 1940 that the didactic impulse was responsible for American ecology's passion for cryptic and awkward terminology:

> Having fought a losing rear-guard action for fifteen years against scholasticism in ecology, I am not surprised to see the growing rigidity of ecological concepts, born of the natural human need for certainty. But I am still very doubtful whether these elaborate systems of relationship, with their uncouth bastard classical names, are in the majority of cases underpinned by a sufficient body of field records.[8]

He was referring to community ecology rather than the new population ecology, but similar fears have been directed toward theoretical ecology and the speed with which textbooks threaten to convert speculation into fact. Ecologists' periodic exhortations to "study nature, not books" can still be heard, echoing Louis Agassiz and T. H. Huxley, whose ideas encouraged budding ecologists a hundred years ago.

These modern exhortations are a response to the perceived ahistorical character of many mathematical approaches, the use of which was often accompanied by a disparagement of history itself in an earlier, optimistic age. MacArthur and others promoted their ideas partly by emphasizing the differences between "science" and "natural history," and by playing upon the impression that history involved little more than endless fact collecting. But this dichotomy caricatures history as mere antiquarianism. In fact, the historian is just as much interested in generalizations and the testing of hypotheses as the scientist, and certainly recognizes the difference between gathering facts and uncovering general patterns.

In ecology the need to balance theory and observation was recognized right from the start, even when it was also admitted that the ideal was not always upheld and that ecologists were in danger of being swamped by the facts. This reciprocal habit of thought was the one advocated by Hutchinson, and it did not depend on the use of mathematical models. What the modern controversy has usefully brought out is the need for biologists explicitly to recognize the special nature of historical questions and perhaps to adopt a method (or philosophy) appropriate to such problems rather than to rely on philosophies generated for deterministic problems encountered in the physical sciences. Ernst Mayr has written on this theme at some length.[9] It raises further questions about biology's changing relation to physics in the early twentieth century, and what that means for the growth of biology as an autonomous science, distinct both from medicine and from physics. These questions deserve more study.

In ecology the success of the mathematical approach, both in generating hypotheses and in raising the status of ecology, is undeniable. But eventually it becomes necessary to pay attention to the historical dimension of these problems of science. It is interesting to observe the change of tone taken by E. O. Wilson, who in collaboration with George Oster wrote another Princeton monograph on optimization theory in evolutionary biology, published in 1978. Their book ended with a critique of the methods presented in the nearly three hundred pages preceding it. They too noticed a parallel between the old argument from design and the modern engineering mentality:

> There is little chance that contemporary biologists will repeat the mistake of the early physicists and see in their abstractions the hand of God. But they are still prone to see His surrogate in the automatic processes leading populations to ever higher levels of adaptation. We will show that not even this relatively modest theme can be translated into mathematically sound arguments.[10]

They concluded later by recognizing the importance of attending to the historical aspects of evolutionary events:

> We must always bear in mind the crucial fact that evolution is a history-dependent process. Adaptations are not "designed" *de novo* by nature. Rather they are jury-rigged using the material available at the time. Evolution, in the words of Jacob (1977), is a "tinkerer," not an engineer! As systems become more complex, the historical accidents play a more and more central role in determining the evolutionary path they will follow.[11]

This critique is only one example of a generally greater awareness of history that recurs in the literature from the mid-1970s on, in response to the mathematical effusions of the previous decade and a half. Oster and Wilson realized that an awareness of history did not imply that everything was random, that there were no patterns, or that a historical discipline was not "scientific." They only meant to suggest that mathematical models had to be used with caution as "provisional guides" to research. This cyclical shift—from a period of optimism in modeling, with an attendant belief that mathematics would raise the status of ecology, to a period of reappraisal characterized by an awareness of the historical dimensions in biology—is characteristic of ecological and evolutionary science.

Other patterns emerge from the preceding histories as well; these reveal the constancies of human behavior and social interactions. Of special interest are the periods of intense polemical dispute which have engaged biologists and mathematicians at regular intervals. These involve only a small minority of a given community, but their intensity inevitably attracts attention. I have recorded three examples of such disputes: the logistic curve debate; W. R. Thompson's debate with the mathematicians in the 1930s; and the disputes at Cold Spring Harbor in 1957. The more recent controversies which began in the 1970s have a similar polemical character.

Although each of these took place in a different context, there are underlying similarities. Most striking is their unusual vehemence, characterized by personal attacks and philosophical arguments of dubious merit. The prevalence of these kinds of debates in science have often been noted in the historical literature. They are usually attributed to a range of causes seen to be acting in conjunction with the actual scientific point at stake: incompatible philosophical commitments; personality factors; or more general differences in social ideology. All of these factors come into play, but it is also important to notice that such debates occur when people are in competition. While arguing about science and philosophy, people are sometimes also arguing over the distribution of scientific resources; over power, prestige, and money. I do not wish to suggest that science or philosophy are irrelevant. They do account for the differences of opinion or worldview, but they do not necessarily account for a debate's vehement tone.

Such debates, with their overlapping layers of science, philosophy, personality, and economics, are very common, but not every scientific debate will contain all these features. When they are exceptionally polemical, however, this is an indication that there is an economic (or ecological) competition also entering into play. We would therefore expect to see such debates in certain special circumstances: as science becomes professionalized; in small, marginal, or new fields; in times of financial crisis. Furthermore, these debates are not resolved by the people who conduct them. It is not simply a matter of exhorting people to collaborate in a civilized way, for the very act of having to defend one's views in the face of attack tends to make the situation more polarized and less flexible. Extreme actions on one side cause opposite and even more extreme reactions on the other side. It is often an outsider, someone not in the competition, who extracts the sound arguments from both sides and presents the reasoned view of the matter.

The frequency with which a third party stands in to settle a problem draws attention to the remarkable consistency with which most people maintain their ideas in a lifetime. Consistency does not necessarily imply rigidity, but it can lead to rigidity. Scientific views, like other parts of a person's worldview, are often formed early in one's career. Although people may modify an opinion from time to time, it is less usual to change one's outlook in any fundamental way. Debates may therefore seem to divert energy into unprofitable seesawing between extreme positions. But they draw attention to problems and perhaps encourage other people to formulate solutions to issues they would not otherwise have considered. The solution to the logistic curve debate was neither Pearl's nor Wilson's, but it was more interesting than the choice either man offered.

The difficulty of breaking out of one's own intellectual mold is also part of the reason that advances are often made by people who have come from a different field, or by collaborative projects. But the success of imported ideas depends on how well the outsider can establish social and institutional security in the new chosen field. I began this book with the story of a relative failure and ended with the story of a success. It is instructive to reflect upon the similarities between Lotka and MacArthur, as well as the differences. Despite their very different goals, they both had much the same point of view. Both saw the organic world in terms of ideal types of organisms playing out their lives according to certain strategies designed to maximize some measure of fitness, though it was not clear just what fitness meant or how this quantity could be maximized. Each tried to formulate new questions connecting the strategies of the individual with the patterns of whole communities. But Lotka, operating too much by himself at a time when ecology was at an embryonic stage and its practitioners' knowledge of mathematics slim, had no audience for the majority of his ideas. More crucially perhaps, he had no biological collaborator to help create an audience. Yet many of the questions and methods he proposed were rediscovered in the 1950s and 1960s, just as his book was posthumously reprinted. It was the attempted union of population genetics with population ecology which brought about this change, yet even if Lotka had not been addressing exactly the same questions, some of his intuitions about what to look for were good.

MacArthur, on the other hand, had the benefit of three decades of theory before him; a more adventurous ecological community, with better mathematical training; Hutchinson's imprimatur; a number of good collaborators; and a shrewder sense of how to make an impact. Certainly he knew his warblers, and that was an important difference, but he also knew how to make use of the opportunities available to him. Even when historical answers began to receive more attention than MacArthur had given them, the problems he thought were interesting remained important to ecologists. But the story of the failure, no less than that of the success, is still valuable for revealing the nature of the formal and informal ties between individuals which characterize modern professional science.

Notes

Chapter 1. Prologue: The Entangled Bank

1. Charles Darwin, *On the Origin of Species. A Facsimile of the First Edition* (Cambridge, Mass. and London: Harvard University Press, 1964), 489.

2. Robert C. Stauffer, ed., *Charles Darwin's Natural Selection; Being the Second Part of His Big Species Book Written from 1856 to 1858* (Cambridge: Cambridge University Press, 1975), 175–78. Here the minimum calculation for elephants is listed as five million.

3. Darwin, *Origin of Species*, 62.

4. Stauffer, *Darwin's Natural Selection*, 208.

5. Darwin, *Origin of Species*, 77. See also Dov Ospovat, *The Development of Darwin's Theory: Natural History, Natural Theology, and Natural Selection, 1838–1859* (Cambridge: Cambridge University Press, 1981), 191–209.

6. Darwin, *Origin of Species*, 78.

7. Robert C. Stauffer, "Haeckel, Darwin, and Ecology," *Quarterly Review of Biology* 32 (1957): 138–44.

8. Eugene Cittadino, "Ecology and the Professionalization of Botany in America, 1890–1905," *Studies in History of Biology* 4 (1980): 171–98. In zoology, the same trends are illustrated in Karl Semper, *Animal Life, As Affected by the Natural Conditions of Existence* (New York: Appleton, 1881).

9. Edward Lurie, *Louis Agassiz; A Life in Science* (Chicago: University of Chicago Press, 1960), 242. See also T. H. Huxley, *Science and Education*, vol. 3, *Collected Essays* (London: Macmillan, 1893).

10. Leland O. Howard, "Stephen Alfred Forbes, 1844–1930," *Biographical Memoirs of the National Academy of Sciences of the U.S.A.* 15 (1932): 1–54.

11. Stephen A. Forbes, "History of the Former State Natural History Societies of Illinois," *Science* 26 (1907): 895.

12. Stephen A. Forbes, "The Food of Birds," *Bulletin of the Illinois State Laboratory of Natural History* 1 (1880): 83. Reprinted in *Ecological Investigations of Stephen Alfred Forbes* (New York: Arno, 1977).

13. Herbert Spencer, *First Principles*, 4th ed., rev. and enl. (New York: Thomas Y. Crowell, 1880).

14. Ibid., sec. 80.

15. Ibid., sec. 60.

16. Ibid., sec. 92.

17. Ibid., sec. 97.

18. Ibid., sec. 96.

19. Ibid., sec. 145.

20. Herbert Spencer, *The Principles of Biology*, 2 vols. rev. and enl. ed. (New York: D. Appleton, 1896), vol. 2, pt. 6.

21. Ibid., sec. 363.

22. Ibid., sec. 364.

23. Ibid., secs. 373–77.

24. Stephen A. Forbes, "On Some Interactions of Organisms," *Bulletin of the Illinois State Laboratory of Natural History* 1 (1880): 1–17. Reprinted in Forbes, *Ecological Investigations*.

25. Forbes, "The Lake As a Microcosm," *Bulletin of the Illinois State Laboratory of Natural History* 15 (1887; reprint 1925): 549–50.

26. Forbes, "Lake as a Microcosm," 550.

27. Cittadino, "Professionalization of Botany," 181.

28. Stephen A. Forbes, *Illinois State Laboratory of Natural History, Semi-Annual Report for 1893–94* (Champaign, Ill.: University of Illinois), 24.

29. Karl Möbius, "The Oyster and Oyster-Culture," *U.S. Commission of Fish and Fisheries, Report for 1880* (Washington, D.C.: 1883). Möbius was the first to propose the term "biocoenosis" for "a community where the sum of species and individuals, being mutually limited and selected under the average external conditions of life, have, by means of transmission, continued in possession of a certain definite territory."

30. Forbes, "Lake As a Microcosm."

31. Frederic E. Clements, *Plant Succession, An Analysis of the Development of Vegetation*, publication no. 242 (Washington, D.C.: Carnegie Institute, 1916).

32. F. E. Clements, J. E. Weaver, and H. C. Hanson, *Plant Competition; An Analysis of Community Functions* (Washington, D.C.: Carnegie Institute, 1929), 10.

33. Garland E. Allen, *Life Science in the Twentieth Century* (Cambridge: Cambridge University Press, 1978).

34. Herbert Spencer Jennings, *Behavior of the Lower Organisms* (1906; reprint, Bloomington: Indiana University Press, 1962), 349.

35. Alfred North Whitehead, *Science and the Modern World* (New York: Macmillan, 1925).

36. Henry C. Cowles, "The Physiographic Ecology of Chicago and Vicinity: A Study of the Origin, Development, and Classification of Plant Societies," *Botanical Gazette* 31 (1901): 73–108; 145–82; idem, "The Causes of Vegetative Cycles," ibid. 51 (1911): 161–83.

37. Cittadino, "Professionalization of Botany," 188.

38. "Charles Christopher Adams," *National Cyclopedia of American Biography* 46 (1963): 258–59; Paul B. Sears, "Charles C. Adams, Ecologist," *Science* 123 (1956): 974.

39. Charles C. Adams, "Migration As a Factor in Evolution: Its Ecological Dynamics," *American Naturalist* 52 (1918): 465–90. Adopting a similar point of view, Arthur G. Tansley introduced the term "ecosystem" in "The Use and Abuse of Vegetational Concepts and Terms," *Ecology* 16 (1935): 284–307. Raymond L. Lindeman later reintroduced this term in a slightly different context in "The Trophic-Dynamic Aspect of Ecology," *Ecology* 23 (1942): 399–418.

40. Cynthia E. Russett, *The Concept of Equilibrium in American Social Thought* (New Haven and London: Yale University Press, 1966).

41. W. D. Bancroft, "A Universal Law," *Science* 33 (1911): 159–79.

42. Lawrence J. Henderson, *The Fitness of the Environment: An Inquiry into the Biological Significance of the Properties of Matter* (New York: Macmillan, 1913). See also Russett, *Equilibrium in American Social Thought*, 111–24.

43. Adams, "Migration As a Factor in Evolution," 469.

44. Thorstein Veblen, "The Evolution of the Scientific Point of View," in *The Place of Science in Civilisation and Other Essays* (New York: Russell and Russell, 1961), 37.

45. Veblen, "The Place of Science in Modern Civilisation," in *The Place of Science*, 17.

46. Adams, "Migration As a Factor in Evolution," 469.

47. Charles C. Adams, "An Outline of the Relations of Animals to Their Inland Environments," *Bulletin of the Illinois State Laboratory of Natural History* 11 (1915): 13.

48. Charles C. Adams, *Guide to the Study of Animal Ecology* (New York: Macmillan, 1913), 80. This annotated bibliography lists several more sources of the process method of ecology.

49. Forbes, "The Food of Birds," 81.

50. Charles C. Adams, "The Relation of General Ecology to Human Ecology," *Ecology* 16 (1935): 316–35.

51. Stephen A. Forbes, *Illinois State Laboratory of Natural History, Semi-Annual Report for 1889–90* (Champaign, Ill.: University of Illinois).

Chapter 2. The World Engine

1. Alfred James Lotka, *Elements of Physical Biology* (Baltimore: Williams and Wilkins, 1925). Reprinted as *Elements of Mathematical Biology* (New York: Dover, 1956).

2. Adams to Lotka, 31 March 1925; Lotka to Adams, 1 April 1925; Lotka to C. C. Thomas (Williams and Wilkins), 29 July 1925, in Alfred James Lotka Papers, Box 4, Princeton University Archives, Princeton, N.J. (hereafter Lotka Papers).

3. Eugene P. Odum, *Fundamentals of Ecology*, 3d ed. (Philadelphia: W. B. Saunders, 1971), 251; idem, "Energy Flow in Ecosystems: A Historical Review," *American Zoologist* 8 (1968): 11–18.

4. Tansley, "Vegetational Concepts and Terms," 284–307; Lindeman, "Trophic-Dynamic Aspect of Ecology," 399–418.

5. Vladimir I. Vernadsky, *La Géochimie* (Paris: Felix Alcan, 1924); *La Biosphère* (Paris: Felix Alcan, 1929).

6. Ludwig von Bertalanffy, *General System Theory: Foundations, Development, Applications*, rev. ed. (New York: George Braziller, 1973).

7. Paul A. Samuelson, *Foundations of Economic Analysis* (Cambridge: Cambridge University Press, 1947); Henry Schultz, *The Theory and Measurement of Demand* (Chicago: University of Chicago Press, 1938); Herbert A. Simon, *Models*

of Man, Social and Rational: Mathematical Essays on Rational Human Behavior in a Social Setting (New York: Wiley, 1957).

8. Herbert A. Simon, review of *Elements of Mathematical Biology*, by A. J. Lotka, *Econometrica* 27 (1959): 493.

9. *Festschrift zur Feier des 500 jährigen Bestehens der Universität Leipzig* (Leipzig: 1909).

10. Wilhelm Ostwald, *Vorlesungen über Naturphilosophie* (Leipzig: 1902).

11. Biographical information is from the following sources: *Population Index* 16 (1950): 22–29; *Journal of the American Statistical Association* 45 (1950): 138–39; *Dictionary of Scientific Biography*, vol. 8; *Who Was Who in America*, vol. 2. The source of information on Lotka's involvement in nitrogen fixation research was a publicity statement prepared by Lotka for the British edition of his book, in the Lotka Papers. No other reference to this research has been found.

12. See the first thirteen entries of the bibliography attached to Lotka, *Elements of Mathematical Biology.* The eleventh entry was not included in his submission.

13. Alfred J. Lotka, *School of Science and Mathematics* 7 (1907): 595. Reprinted in *Scientific American Supplement* 73 (1912).

14. Lotka, *Proceedings of the National Academy of Sciences of the U. S. A.* 6 (1920): 410–15.

15. Pearl to Lotka, 18 April 1921, A. J. Lotka file, Raymond Pearl Papers, American Philosophical Society Archives, Philadelphia (hereafter Pearl Papers).

16. Lotka to Pearl, 20 April 1921, A. J. Lotka file, Pearl Papers.

17. Pearl to Lotka, 25 July 1921, A. J. Lotka file, Pearl Papers.

18. Pearl to Lotka, 18 May 1921, A. J. Lotka file, Pearl Papers.

19. Lotka to Pearl, 1 June 1921, A. J. Lotka file, Pearl Papers.

20. Lotka to Pearl, 28 June 1921, A. J. Lotka file, Pearl Papers.

21. Lotka to Pearl, 7 November 1921, A. J. Lotka file, Pearl Papers.

22. Pearl to Lotka, 9 November 1921, A. J. Lotka file, Pearl Papers.

23. Lotka to Pearl, 21 November 1921, A. J. Lotka file, Pearl Papers.

24. Lotka to Pearl, 8 January 1922, A. J. Lotka file, Pearl Papers. I have been unable to discover how Lotka supported himself from 1919 until 1922. In late 1921 his letters were addressed from Scranton, Pennsylvania, and the figure of $3,000 cited by Lotka refers to whatever employment he had there.

25. Pearl to Lotka, 11 January 1922, A. J. Lotka file, Pearl Papers.

26. Pearl to Lotka, 27 January 1922, A. J. Lotka file, Pearl Papers.

27. William E. Ritter and Edna W. Bailey, "The Organismal Conception: Its Place in Science and Its Bearing on Philosophy," *University of California Publications in Zoology* 31 (1927–29): 307–58.

28. *The Unconscious: A Symposium* (Freeport, N.Y.: Books for Libraries Press, 1928); Jan C. Smuts, *Holism and Evolution* (New York: Macmillan, 1926); Donna Haraway, *Crystals, Fabrics, and Fields: Metaphors of Organicism in Twentieth-Century Developmental Biology* (New Haven and London: Yale University Press, 1976).

29. D'Arcy Wentworth Thompson, *On Growth and Form*, 2 vols., (Cambridge: Cambridge University Press, 1917).

30. Alexander Forbes, "Biophysics," *Science* 52 (1920): 331–32.

31. David Burns, *An Introduction to Biophysics* (London: J. and A. Churchill, 1921).

32. Lotka to Pearl, 31 January 1922, A. J. Lotka file, Pearl Papers.

33. Lotka, *Elements*, 184.

34. Lotka to Ernest Merritt, physicist, 1 March 1927, Box 4, Lotka Papers.

35. W. Porstmann, "Ein Problem aus der physikalischer Zoologie: Einfluss physikalischer Momente auf die Gestalt der Fische," *Prometheus* 26 (1915): 267–70; 284–86; 300–303.

36. Ostwald, *Vorlesungen*, 342–44.

37. Alfred J. Lotka, "The Law of Evolution As a Maximal Principle," *Human Biology* 17 (1945), 176n.

38. Lotka, *Elements*, 13–16.

39. Lotka, *Elements*, 154–55. This comparison is first made in Alfred J. Lotka, "Studies on the Mode of Growth of Material Aggregates," *American Journal of Science*, 4th ser., 24 (1907): 199–216.

40. Alfred J. Lotka, "Evolution in Discontinuous Systems," *Journal of the Washington Academy of Sciences* 2 (1912): 2–6; 49–59; 66–74. See also idem, "Evolution and Irreversibility," *Science Progress* 14 (1920): 406–17; idem, *Elements*, 45.

41. Lotka, *Elements*, 158.

42. Ibid., 24.

43. Jean Perrin, *Traité de chimie physique: les principes* (Paris: 1903), 142–43.

44. Alfred J. Lotka, "Natural Selection As a Physical Principle," *Proceedings of the National Academy of Sciences of the U. S. A.* 8 (1922): 151–54.

45. Ibid., 152–53.

46. Ibid., 152.

47. Ludwig Boltzmann, *Der zweite Hauptsatz der mechanischen Wärmetheorie* (Vienna: Gerold, 1886); idem, "The Second Law of Thermodynamics," *Theoretical Physics and Philosophical Problems; Selected Writings* (Dordrecht and Boston: D. Reidel, 1974).

48. Alfred J. Lotka, "Contribution to the Energetics of Evolution," *Proceedings of the National Academy of Sciences of the U. S. A.* 8 (1922), 147–51.

49. Ibid.

50. Lotka, *Elements*, 357–58.

51. Lotka, "Contribution to the Energetics," 147–48.

52. Ibid., 149. See also Lotka, *Elements*, 357.

53. Alfred J. Lotka, "Note on Moving Equilibria," *Proceedings of the National Academy of Sciences of the U.S.A.* 7 (1921): 171–72; idem, "Contribution to the Energetics," 149; idem, *Elements*, 334–35.

54. The third law of thermodynamics on the unattainability of absolute zero had just been formulated as an extension of the 1906 heat theorem of Hermann Walther Nernst.

55. Wilhelm Ostwald, *Natural Philosophy* (New York: Henry Holt, 1910), 171.

56. Sadi Carnot, *Réflexions sur la puissance du feu* (1824).

57. Alfred J. Lotka, "Evolution from the Standpoint of Physics," *Scientific American Supplement* 75 (1913): 345.

58. Spencer, *First Principles*, sec. 188.

59. C. Gide and C. Rist, *A History of Economic Doctrines*, 2d ed. (London: George G. Harrap, 1948), 499–514.

60. William Stanley Jevons, *The Theory of Political Economy*, 2d ed., rev. (London: Macmillan, 1879). Lotka used the 1911 edition, which has the same text. Also, Vilfredo Pareto, *Manual of Political Economy*, trans. A. S. Schwier (New York: Augustus M. Kelley, 1971).

61. Jevons, *Political Economy*, vii.

62. Herbert Spencer, *Principles of Psychology*, 3d ed. (New York: D. Appleton, 1902), secs. 124–25.

63. Alfred J. Lotka, "An Objective Standard of Value Derived from the Principle of Evolution," *Journal of the Washington Academy of Sciences* 4 (1914): 409–18; 447–57; 499–500.

64. Alfred J. Lotka, "Efficiency As a Factor in Organic Evolution," *Journal of the Washington Academy of Sciences* 5 (1915): 360–68; 397–403.

65. Lotka, "Objective Standard of Value," 416.

66. Lotka, "Efficiency as a Factor," 360–68, 397–403.

67. Lotka, *Elements*, 330–57.

68. Ibid., 350–53.

69. For a review of economic thinking in recent ecology, see David J. Rapport and James E. Turner, "Economic Models in Ecology," *Science* 195 (1977): 367–73.

70. Lotka, "Studies on the Mode of Growth," 199–216.

71. Lotka, "Evolution from the Standpoint of Physics," 345.

72. An outline for the projected book and part of a manuscript, both undated, are in the Lotka Papers.

73. Manuscript of projected book, chap. 1, Lotka Papers.

74. Herbert Spencer, "The Social Organism," in *Essays, Scientific, Political and Speculative* (New York: D. Appleton, 1891), vol. 1, 864.

75. Wilhelm Ostwald, "Machines and Living Creatures," *Scientific American Supplement* 70 (1910): 55; idem, "The System of the Sciences," *Book of the Opening of the Rice Institute* (Houston: 1912), vol. 3, 864.

76. This idea is mainly discussed in the concluding chapter of the *Elements*.

77. Lotka, *Elements*, 387.

78. Alfred J. Lotka, "Biassed Evolution," *Harper's Magazine* 148 (May 1924): 755–66.

79. Frederick Jackson Turner, "Social Forces in American History," in *The Frontier in American History* (1920; reprint, New York: Holt, Rinehart and Winston, 1962); Edward A. Ross, *Social Control* (1901; reprint, New York and London: Johnson Reprint, 1970); Herbert Croly, *The Promise of American Life* (1909; reprint, Cambridge, Mass.: Belknap Press, 1965). For a discussion of relevant themes in social thought, see Morton White, *Social Thought in America; The Revolt against Formalism* (Boston: Beacon Press, 1957); Russett, *Equilibrium in American Social Thought.*

80. W. T. Lhamon, Jr., "Horatio Alger and American Modernism," *American Studies* 17 (1976): 11–27.

81. Brooks Adams, "The New Industrial Revolution," *Atlantic Monthly* 87 (1901): 165.

82. H. G. Wells, *The Future in America: A Search after Realities* (London: Chapman and Hall, 1906), 70–71.

83. Lotka, *Elements*, 417–34.

84. Undated circular from Williams and Wilkins, "A Catalogue of Good Reading," Lotka Papers.

85. Copies of letters accompanying the books sent out for review indicate that none went to ecologists, Lotka Papers.

86. Lotka to C. C. Thomas, 17 August 1926, Lotka Papers.

87. Lotka to Ernest Merritt, 1 March 1927, Box 4, Lotka Papers.

88. Edwin B. Wilson, review of *Elements of Physical Biology* by A. J. Lotka, *Science* 66 (1927): 281–82. *Die Ausdehnungslehre* by Hermann Grassmann, published in 1844, introduced a new branch of mathematics dealing with the science of pure extension. His work was not appreciated until the end of his life, after Riemann and Helmholtz had made similar investigations into the theory of extension.

89. Lotka to C. C. Thomas, 17 August 1926, Box 4, Lotka Papers. The only review he liked was by W. A. White, *Psychoanalytic Review* 12 (1925): 323–30. Other reviews, on the whole favorable, were: E. W. Kopf, *Journal of the American Statistical Association* 20 (1925): 452–56; G. Bohn, *Revue générale des sciences pures et appliquées* 37 (1926): 217–18; G. Brunelli, *Rivista di Biologica* 8 (1926): 103; anonymous reviews in *Science Progress* 20 (1925): 337–39; *Nature* 116 (1925): 461.

90. R. S. Gill (Williams and Wilkins) to Lotka, 30 December 1935, Lotka Papers.

91. R. S. Gill to Lotka, 29 February 1940, Lotka Papers.

92. Frank Lorimer, "The Development of Demography," in P. M. Hauser and O. D. Duncan, eds., *The Study of Population: An Inventory and Appraisal* (Chicago and London: University of Chicago Press, 1959), 156. For an assessment of Lotka's work as a demographer, see, in the same volume, R. B. Vance, "The Development and Status of American Demography," 286–313. See also Alvaro Lopez, *Problems in Stable Population Theory* (Princeton, N.J.: Office of Population Research, 1961).

Chapter 3. The Quantity of Life

1. Leland O. Howard, *A History of Applied Entomology*, vol. 84, Miscellaneous Collection (Washington, D.C.: Smithsonian Institution, 1931), 164; idem, *Fighting the Insects; The Story of an Entomologist* (New York: Macmillan, 1933); Paul DeBach, *Biological Control by Natural Enemies* (Cambridge: Cambridge University Press, 1974).

2. R. L. Doutt, "Vice, Virtue and the Vedalia," *Bulletin of the Entomological Society of America* 4 (1958): 119–23.

3. Leland O. Howard, "Entomology and the War," *Scientific Monthly* 8 (1919): 109–17.

4. Samuel A. Graham, "Forest Insect Populations," *Ecological Monographs* 9 (1939): 301–10.

5. Harry S. Smith, "The Present Status of Biological Control Work in California," *Journal of Economic Entomology* 19 (1926): 294–302.

6. Howard, *History of Applied Entomology*, 505.

7. William R. Thompson, "Biological Control and the Theories of the Interactions of Populations," *Parasitology* 31 (1939): 299–388.

8. Charles S. Elton, "Periodic Fluctuations in the Numbers of Animals: Their Causes and Effects," *Journal of Experimental Biology* (British) 2 (1924–25): 119–63.

9. Ibid.

10. Charles Elton, *Animal Ecology* (London: Sidgwick and Jackson, 1927). See Huxley's preface.

11. There were naturalists, such as F. B. Sumner, who were interested in evolution and whose work included ecological studies. But I base my judgment here on the discipline of ecology as defined in the early textbooks of the 1920s. The discipline at that time was no more concerned with evolution than sociology is concerned with human evolution. Thus although the study of evolution necessarily involves the ecology of organisms, the discipline of ecology does not necessarily include the study of evolution. The following ecological texts do not mention evolution or natural selection in their indices: W. B. McDougall, *Plant Ecology* (Philadelphia: Lea & Febiger, 1927); J. E. Weaver and F. E. Clements, *Plant Ecology* (New York: McGraw-Hill, 1929); V. E. Shelford, *Laboratory and Field Ecology: The Responses of Animals as Indicators of Correct Working Methods* (Baltimore: Williams and Wilkins, 1929); F. E. Clements and V. E. Shelford, *Bio-ecology* (New York: Wiley, 1939); Royal N. Chapman, *Animal Ecology, With Especial Reference to Insects* (New York and London: McGraw-Hill, 1931). A later text which includes only six pages on evolution is A. S. Pearse, *Animal Ecology*, 2d ed. (New York and London: McGraw-Hill, 1939).

12. Elton, *Animal Ecology*, 179.

13. R. A. Fisher, *The Genetical Theory of Natural Selection* (Oxford: Oxford University Press, 1930), 44.

14. G. C. Robson and O. W. Richards, *The Variation of Animals in Nature* (London: Longmans, Green, 1936), 352–55.

15. Elton, "Periodic Fluctuations."

16. Ibid., 160.

17. Elton, *Animal Ecology*, 187.

18. L. L. Woodruff, "Observations on the Origin and Sequence of the Protozoan Fauna of Hay Infusions," *Journal of Experimental Zoology* 12 (1912): 205–64. See also G. Evelyn Hutchinson, "Lorande Loss Woodruff, 1879–1947," *Biographical Memoirs of the National Academy of Sciences of the U. S. A.* 52 (1980): 471–85.

19. W. C. Allee, *Animal Aggregations: A Study in General Sociology* (Chicago: University of Chicago Press, 1931).

20. R. N. Chapman, "The Confused Flour Beetle," *17th Minnesota State Entomology Report*, 1918, 73–94; Samuel A. Graham, "Royal Norton Chapman, 1889–1939," *Annals of the Entomological Society of America* 34 (1941): 521–24.

21. Raymond Pearl, "The Trends of Modern Biology," *Science* 56 (1922): 581–92.

22. H. S. Jennings, "Raymond Pearl, 1879–1940," *Biographical Memoirs of the National Academy of Sciences of the U. S. A.* 22 (1942): 295–347.

23. Raymond Pearl, *To Begin With; Being Prophalaxis against Pedantry* (New York: Alfred A. Knopf, 1927).

24. Karl Pearson, *The Grammar of Science*, 2d ed. (London: Adam & Charles Black, 1900), 31. See the first three chapters for Pearson's views on scientific method.

25. Karl Pearson, *The Chances of Death, And Other Studies in Evolution*, 2 vols. (London: Edward Arnold, 1897). See essays in vol. 1, "The Chances of Death," "Reproductive Selection," and "Socialism and Natural Selection."

26. Karl Pearson, "Darwinism, Biometry and Some Recent Biology," *Biometrika* 7 (1910): 368–85.

27. William B. Provine, *The Origins of Theoretical Population Genetics* (Chicago and London: University of Chicago Press, 1971), 92–108.

28. Raymond Pearl, "An Appeal," *Science* 50 (1919): 524–25; Pearl to E. M. East, 30 November 1919, E. M. East file #1, Pearl Papers.

29. Raymond Pearl, *The Rate of Living; Being an Account of Some Experimental Studies on the Biology of Life Duration* (New York: Alfred A. Knopf, 1928), 2.

30. Ibid., 3.

31. Ibid., 93.

32. Pearl to W. M. Wheeler, 9 May 1923, W. M. Wheeler file, Pearl Papers.

33. Pearl to L. J. Henderson, 16 May 1923, L. J. Henderson file, Pearl Papers.

34. Raymond Pearl, "A Plan for Research on the Biology of Life Duration and Extension." Cover letter is dated 25 March 1924. Record Group 1.1, Series 200D, Box 145, Folder 1791, Rockefeller Foundation Archives, North Tarrytown, N.Y. (hereafter RF Archives).

35. E. R. Embree to Pearl, 25 May 1925, Record Group 1.1, 200D, Box 145, Folder 1791, RF Archives.

36. See correspondence and documents in Boxes 145–46, Record Group 1.1, 200D, Johns Hopkins University—Biological Research, 1924–1939, RF Archives.

37. Pearl to Greenwood, 21 May 1925, Major Greenwood file, Pearl Papers.

38. Pearl to Greenwood, 15 June 1925, Major Greenwood file, Pearl Papers.

39. A yearly summary of staff, activities, and publications of the institute is in "Report of the Director of the Institute for Biological Research," *Annual Report of the Johns Hopkins University*, 1925–1926 to 1929–1930.

40. Raymond Pearl, "The Biology of Death. VII. Natural Death, Public Health, and the Population Problem," *Scientific Monthly* 13 (1921), 212; idem, *The Biology of Population Growth* (New York: Alfred A. Knopf, 1925), 2. On the causes of the war and population growth, see E. D. Durand, "Some Problems of Population Growth," *Journal of the American Statistical Association* 15 (1916): 129–48; Harold Cox, *The Problem of Population* (New York and London: G. P. Putnam, 1923), 77–108; John Holland Rose, *The Origins of the War* (Cambridge: University Press, 1914), 47–48. Alex M. Carr-Saunders argued against this view, which he acknowledged to be widespread, in *The Population Problem: A Study in Human Evolution* (Oxford: Clarendon, 1922), 305–7.

41. Raymond Pearl, *The Nation's Food* (Philadelphia: W. B. Saunders, 1920), 17.

42. Raymond Pearl, "The Effect of War on the Chief Factors of Population Change," *Science* 51 (1921): 553–56.

43. Raymond Pearl and Lowell J. Reed, "On the Rate of Growth of the Population of the United States Since 1790 and Its Mathematical Representation," *Proceedings of the National Academy of Sciences of the U. S. A.* 6 (1920): 275–88.

44. L. A. J. Quetelet, *Sur L'homme et le développement de ses facultés: ou Essai de physique sociale*, 2 vols. (Paris, 1835), vol. 2, 277.

45. Pierre-François Verhulst, "Notice sur la loi que la population suit dans son accroissement," *Correspondances mathématiques et physiques* 10 (1938): 113–21; idem, "Recherches mathématiques sur la loi d'accroissement de la population," *Nouveaux Mémoires de l'Académie Royale des Sciences et Belles-Lettres de Bruxelles*, ser. 2, 18 (1845): 3–38. For more information on Verhulst, see G. Evelyn Hutchinson, *An Introduction to Population Ecology* (New Haven and London: Yale University Press, 1978), chap. 1.

46. Verhulst, "Recherches mathématiques," 8.

47. The distinction is between calculation to arrive at a numerical answer and, for example, theory of proportion. See "Logistique" in Larousse, *Grand Dictionnaire Universel du XIXe Siècle*, 1865; *Dictionnaire Encyclopédique Quillet.*

48. Verhulst, "Recherches mathématiques," 18.

49. Verhulst, "Deuxième mémoire sur la loi d'accroissement de la population," *Mémoires de l'Académie Royale des Sciences, des Lettres et des Beaux-Arts de Belgique*, ser. 2, 20 (1847), 6.

50. Quetelet's opinion is in an obituary notice of Verhulst, translated into English by John R. Miner, in *Human Biology* 5 (1933): 673–89.

51. Pearl does not say how he learned of Verhulst's work, but his first reference to it is in Pearl, "Biology of Death," 206n.

52. T. B. Robertson, "On the Normal Rate of Growth of an Individual and Its Biological Significance," *Archiv für Entwicklungsmechanik der Organismen* 25 (1908): 581–613; idem, "Further Remarks on the Normal Rate of Growth of an Individual and Its Biological Significance," ibid, 26 (1908): 108–18.

53. P. J. Lloyd, "American, German, and British Antecedents to Pearl and Reed's Logistic Curve," *Population Studies* 21 (1967): 99–108.

54. T. B. Robertson, "On the Nature of the Autocatalyst of Growth," *Archiv für Entwicklungsmechanik der Organismen* 37 (1913): 497–508.

55. T. B. Robertson, *The Chemical Basis of Growth and Senescence* (Philadelphia and London: Lippincott, 1923).

56. Raymond Pearl, "Some Recent Studies on Growth," *American Naturalist* 43 (1909): 302–16.

57. Raymond Pearl, *Variation and Differentiation in Ceratophyllum*, Publication no. 58 (Washington, D.C.: Carnegie Institute, 1907), 8.

58. Pearl, "Recent Studies on Growth," 315.

59. Pearl and Reed, "On the Rate of Growth of the Population of the United States," 281.

60. Pearl, *The Biology of Population Growth*, 22.

61. G. Udny Yule, "The Growth of Population and the Factors Which Control

It," *Journal of the Royal Statistical Society of London*, ser. A, 88 (1925): 1–58. See p. 4.

62. Raymond Pearl and Lowell J. Reed, "On the Mathematical Theory of Population Growth," *Metron* 3 (1923): 6–19.

63. A. D. Darbishire, *An Introduction to a Biology* (London: Cassell, 1917), 169.

64. Raymond Pearl, *Studies in Human Biology* (Baltimore: Williams and Wilkins, 1924), 585.

65. Pearl and Reed, "On the Rate of Growth of the Population of the United States," 284–85.

66. Pearl and Reed, "On the Mathematical Theory of Population Growth."

67. Pearl, "Recent Studies on Growth," 310.

68. Pearl and Reed, "On the Rate of Growth of the Population of the United States," 287.

69. Raymond Pearl and Lowell J. Reed, "Skew-growth Curves," *Proceedings of the National Academy of Sciences of the U. S. A.* 11 (1925): 16–22.

70. Pearl and Reed, "On the Mathematical Theory of Population Growth," 6–19. Notation has been changed to conform to modern style.

71. Raymond Pearl, *Introduction to Medical Biometry and Statistics* (Philadelphia and London: W. B. Saunders, 1923), 333.

72. Pearl and Reed, "On the Mathematical Theory of Population Growth," 6.

73. Margaret Sanger, ed., *Proceedings of the World Population Conference, Geneva, 1927* (London: Edward Arnold, 1927). See Pearl's comments in the discussion following his paper, p. 55.

74. Raymond Pearl and Lowell J. Reed, "The Growth of Human Population," in Pearl, *Studies in Human Biology*, 584–637.

75. Pearl later found that the 1920 and 1930 census results did fit the same curve for the United States, but for the 1940 census he refitted the data to a different curve which had a much lower limit of 184 million. This note appeared a week after Pearl's death, and it concluded that either of the two curves fitted the observations equally well. R. Pearl, L. J. Reed, and J. F. Kish, "The Logistic Curve and the Census Count of 1940," *Science* 92 (1940): 486–88.

76. Raymond Pearl, "The Curve of Population Growth," *Proceedings of the American Philosophical Society* 63 (1924): 10–17.

77. Pearl, *Studies in Human Biology*, 596.

78. H. S. Reed, "The Nature of Growth," *American Naturalist* 58 (1924): 337–49.

79. A. G. McKendrick and M. Kesava Pai, "The Rate of Multiplication of Micro-organisms: A Mathematical Study," *Proceedings of the Royal Society of Edinburgh* 31 (1911): 649–55; T. Carlson, "Uber Geschwindigkeit und Grösse der Hefevermehrung in Wurze," *Biochemisches Zeitschrift* 57 (1913): 313–34.

80. Raymond Pearl, "The Biology of Population Growth," *American Mercury* 3 (1924): 293–305; idem, *Biology of Population Growth*, chap. 2.

81. Pearl, "The Biology of Population Growth," 302.

82, Raymond Pearl, "The Growth of Populations," *Quarterly Review of Biology* 2 (1927), 533.

83. For other articles on the logistic curve, see the bibliography in H. S.

Jennings, "Raymond Pearl," *Biographical Memoirs of the National Academy of Sciences of the U. S. A.* 22 (1942): 295–347.

84. "Meeting on the Problem of Forecasting City Populations with Special Reference to New York City," *Journal of the American Statistical Association* 20 (1925): 569–73.

85. Pearl to W. T. Howard, 2 November 1925, W. T. Howard file #2, Pearl Papers.

86. *Report of the British Association for the Advancement of Science*, meeting in Toronto, 1924 (London, 1925), 410, 466.

87. Yule, "The Growth of Population"; see discussion of logistic curve following paper by T. H. C. Stevenson, *Journal of the Royal Statistical Society of London*, ser. A, 88 (1925): 63–90.

88. Sanger, ed., *World Population Conference*, 22–38; 39–58; also Pearl, "Growth of Populations," 532–48.

89. *The Biology of Death* was translated into Swedish in 1924; *The Biology of Population Growth* into Polish in 1925.

Chapter 4. Much Ado

1. See, for instance: W. F. Willcox, "Population and the World War," *Journal of the American Statistical Association* 18 (1922–23): 699–712; G. C. Whipple, review of *The Biology of Death* by R. Pearl, ibid. 18 (1922–23): 926–28; H. M. Flinn, review of *Predicted Growth of New York and Environs* by Raymond Pearl and L. J. Reed, ibid. 19 (1924): 111–13; T. H. C. Stevenson, "The Laws Governing Population," *Journal of the Royal Statistical Society of London*, ser. A, 88 (1925), 68; A. L. Bowley, comments in ibid, ser. A, 88 (1925), 76; H. Hotelling, "Differential Equations Subject to Error, and Population Estimates," *Journal of the American Statistical Association* 22 (1927): 283–314.

2. Pearl, *Studies in Human Biology*, 586–87.

3. A. B. Wolfe, "Is There a Biological Law of Human Population Growth?" *Quarterly Journal of Economics* 41 (1927): 573.

4. Pearl, *Studies in Human Biology*, 587.

5. Wolfe, "Is There a Biological Law?" 576.

6. George H. Knibbs, "A Mathematical Theory of Populations, of Its Character and Fluctuations, and of the Factors Which Influence Them," *Census of the Commonwealth of Australia for 1911*, vol. 1, appendix A (1917), 129.

7. George H. Knibbs, "The Growth of Human Populations, and the Laws of Their Increase," *Metron* 5(3) (1925): 147–62; idem, "The Laws of Growth of a Population," *Journal of the American Statistical Association* 21 (1926): 381–98; Part 2, 22 (1927): 49–59.

8. George H. Knibbs, "The New Malthusianism in the Light of Actual World Problems of Population," *Scientia* 40 (1926): 379–88.

9. Raymond Pearl, "The Biology of Superiority," *American Mercury* 12 (1927): 257–66; idem, "Differential Fertility," *Quarterly Review of Biology* 2 (1927): 102–18. See criticisms by D. G. Paterson and E. G. Williamson, "Raymond Pearl on the Doctrine of 'Like Produces Like,'" *American Naturalist* 63 (1929): 265–73.

10. Raymond Pearl, "Interim Report of the First General Assembly of the International Union for the Scientific Investigation of Population Problems," *Journal of the American Statistical Association* 23 (1928): 306–17.

11. Stevenson, "Laws Governing Population," 75. Lancelot Hogben, "Some Biological Aspects of the Population Problem," *Biological Reviews* 6 (1931): 163–80.

12. Pearl, *Biology of Population Growth*, 212. When Pearl later studied the effectiveness of contraception, he first concluded that birth control did not in fact result in lower fertility among women who used it. Further study showed he had calculated the pregnancy rate incorrectly. He reviewed his work and concluded that birth control did effectively lower fertility, expecially in women of the higher economic classes. Raymond Pearl, "Preliminary Notes on a Cooperative Investigation of Family Limitation," *Milbank Memorial Fund Quarterly* 11 (1933): 37–60; idem, "Second Progress Report on a Study of Family Limitation," ibid. 12 (1934): 248–69.

13. Raymond Pearl, "A Further Note on War and Population," *Science* 53 (1921): 120–21.

14. Pearl, "Growth of Populations," 541.

15. Ibid.

16. Pearl, *Variation and Differentiation in Ceratophyllum.*

17. Pearl, "Growth of Populations," 541.

18. John Brownlee, "Notes on the Biology of a Life-Table," *Journal of the Royal Statistical Society of London* 82 (1919); 34–65: idem, "Density and Death Rate: Farr's Law," ibid. 83 (1920): 280–83. These are discussed in Raymond Pearl and Sylvia L. Parker, "Experimental Studies on the Duration of Life. IV. Data on the Influence of Density of Population on Duration of Life in *Drosophila*," *American Naturalist* 56 (1922): 312–21.

19. Pearl, "Growth of Populations," 532–48; idem, *Biology of Population Growth*, 131–57.

20. Pearl, *Biology of Population Growth*, 257.

21. Ibid., 146–56.

22. Wolfe, "Is There a Biological Law?" 592.

23. Hogben, "Some Biological Aspects of the Population Problem," 167. See also O. W. Richards, "Potentially Unlimited Multiplication of Yeast with Constant Environment, and the Limiting of Growth by Changing Environment," *Journal of General Physiology* 11 (1927–28): 525–38.

24. Raymond Pearl, "The Influence of Density of Population upon Egg Production in *Drosophila melanogaster*," *Journal of Experimental Zoology* 63 (1932): 57–84.

25. Alfred J. Lotka, "Contribution to the Theory of Periodic Reactions," *Journal of Physical Chemistry* 14 (1910): 271–74.

26. Lotka, "Studies on the Mode of Growth," 199–216.

27. L. J. Reed to M. Spiegelman, 31 January 1950, courtesy of Dr. Thomas Holzmann, Office of Population Research, Princeton University.

28. Ibid.

29. Ibid.

30. Lotka, *Elements*, 64–66.

31. Wolfgang Ostwald, *Uber die zeitlichen Eigenschaften der Entwicklungsvorgänge* (Leipzig: 1908). Wolfgang was the son of Wilhelm Ostwald.

32. Alfred J. Lotka, reported in *Journal of the American Statistical Association* 21 (1925): 569–70.

33. Ibid., 569.

34. Alfred J. Lotka, "The Size of American Families in the Eighteenth Century; And the Significance of the Empirical Constants in the Pearl-Reed Law of Population Growth," *Journal of the American Statistical Association* 22 (1927): 154–70.

35. Louis I. Dublin and Alfred J. Lotka, "On the True Rate of Natural Increase of a Population," *Journal of the American Statistical Association* 20 (1925): 305–39.

36. Reported in *Journal of the American Statistical Association* 26 (1931): 194–95. See also Alfred J. Lotka, "Biometric Functions in a Population Growing in Accordance with a Prescribed Law," *Proceedings of the National Academy of Sciences of the U. S. A.* 15 (1929): 793–98.

37. Alfred J. Lotka, "The Structure of a Growing Population," in *Problems of Population*, G. H. L. F. Pitt-Rivers, ed. (London: George Allen and Unwin, 1932). Published in *Human Biology* 3 (1931): 459–93.

38. R. C. Punnett, "Early Days of Genetics," *Heredity* 4 (1950): 1–10.

39. Raymond Pearl, review of *The Social Life of Animals*, by W. C. Allee, *Ecology* 20 (1939), 309.

40. Edwin B. Wilson, "The Statistical Significance of Experimental Data," *Science* 58 (1923): 93–100.

41. Edwin B. Wilson, "Statistical Inference," *Science* 63 (1926): 290–91. See also idem, "Empiricism and Rationalism," ibid. 64 (1926): 47–57.

42. Edwin B. Wilson, "The Population of Canada," *Science* 61 (1925): 87–89.

43. Edwin B. Wilson and W. J. Luyten, "The Population of New York City and Its Environs," *Proceedings of the National Academy of Sciences of the U. S. A.* 11 (1925): 137–43.

44. Raymond Pearl and Lowell J. Reed, "On the Summation of Logistic Curves," *Journal of the Royal Statistical Society of London* 90 (1927): 729–46.

45. "A statistician of the hard-headed English type, who walks in the way of Tooke and Newmarch and Giffen, applying ascertained facts to important practical problems." Quotation from F. Y. Edgeworth, *Economic Journal* 34 (1924): 430.

46. Major Greenwood to Pearl, 12 March 1925, Major Greenwood file, Pearl Papers.

47. *Journal of the Royal Statistical Society of London*, 88 (1925); Bowley's comments p. 79; Yule's reply p. 89.

48. Edwin B. Wilson, "The Logistic or Autocatalytic Grid," *Proceedings of the National Academy of Sciences of the U. S. A.* 11 (1925): 451–56.

49. Pearl to Major Greenwood, 30 March 1925, Major Greenwood file, Pearl Papers. Pearl explained that Wilson's opposition had begun when Wilson had come to Pearl with a request for Pearl to furnish him with problems. Pearl had turned him down, claiming not to want to create bad feeling between them later on, as he did not feel Wilson capable of handling biological problems very well. He also

felt that Wilson was maneuvering him into a position of less influence at the National Academy of Sciences, but the motivations are unclear. What is clear is that Wilson had an intense dislike of Pearl's style, which he felt was inappropriate for a scientist. He was generally known to have a "sharp and critical intelligence" and a "caustic tongue," which came into play in other disputes besides this one with Pearl. See his biography in *Biographical Memoirs of the National Academy of Sciences* 43 (1973).

50. Pearl and Reed, "On the Summation of Logistic Curves," 729–46.

51. R. Pearl, A. C. Sutton, and W. T. Howard, "Experimental Treatment of Cancer with Tuberculin," *Lancet* 216 (1929): 1078–80.

52. R. Pearl, "Proposal for the Continuation of the Institute for Biological Research, with a Program for Its Future Development," 1 February 1929, Folder 1793, Box 145, Record Group 1.1, Series 200D, RF Archives.

53. L. J. Henderson to Pearl, 5 February 1929; Pearl to Henderson, 9 February 1929, in L. J. Henderson file, Pearl Papers.

54. Pearl sent a confidential memorandum to his staff, dated 23 March 1929, announcing his acceptance of the appointment as Professor of Biology at Harvard, to begin 1 July 1930. In Institute for Biological Research file, Pearl Papers.

55. W. M. Wheeler to Pearl, 18 June 1929, in W. M. Wheeler file, Pearl Papers. This file also contains copies of letters sent by Wilson to Henry James, President Lowell, W. M. Wheeler, and members of the Harvard Cancer Commission, all critical of Pearl. Other letters related to the Harvard scandal are in the files of William Henry Welch, Thomas Barbour, L. J. Henderson, E. R. Embree, in the Pearl Papers. See also M. A. Evans and H. E. Evans, *William Morton Wheeler, Biologist* (Cambridge, Mass.: Harvard University Press, 1970), 245ff.; Carl Bode, ed., *The New Mencken Letters* (New York: Dial Press, 1977), 235–37.

56. Some of these are in the files of Record Group 1.1, Series 200D, RF Archives.

57. L. J. Henderson to Pearl, 25 September 1929, in L. J. Henderson file, Pearl Papers.

58. Notes from Max Mason's diary, on a visit to Baltimore on 3 September 1929, to discuss Pearl's situation with President Ames, Folder 1793, Box 145, Record Group 1.1, Series 200D, RF Archives.

59. Notes made after interview between Warren Weaver and Max Mason, 6 May 1938, reviewing the events following the Harvard episode, Folder 1795, Box 145, Record Group 1.1, Series 200D, RF Archives.

60. H. S. Jennings to Pearl, 29 November 1929, in Institute for Biological Research file, Pearl Papers.

61. H. S. Jennings, "Report of the Zoological Laboratory," *Johns Hopkins University Circular*, 1924–1925.

62. Rockefeller Foundation to Joseph Ames, 14 February 1930, in Joseph S. Ames file, Pearl Papers.

63. His department was given a budget of $31,000 per year, plus $2,500 each from the School of Hygiene and the Medical School. Pearl's salary remained at $15,000. Joseph S. Ames to Pearl, 17 April 1930, in Joseph S. Ames File, Pearl Papers.

64. Although the university was supposed to make up the decreases in the

Rockefeller contributions, it requested in 1933 and subsequent years to have its contributions reduced, so that the actual amount available for biological research decreased each year. Records in the Rockefeller Foundation Archives indicate that Pearl's allotment remained constant, leaving the other departments to absorb the decreases. An excerpt from a letter by S. O. Mast of the Zoology Department, 17 March 1936, mentions ill-feeling over the fact that 67 percent of the Rockefeller funds went to the School of Hygiene (Pearl's laboratory) while the rest went to the other biology departments. A summary of the foundation's support of biological research at Johns Hopkins is in Record Group 1.1, Series 200D, RF Archives.

65. E. B. Wilson and H. C. Maher, "Cancer and Tuberculosis with Some Comments on Cancer and Other Diseases," *American Journal of Cancer* 16 (1932): 227–50.

66. Edwin B. Wilson, "The Value of Statistical Studies of the Cancer Problem," *American Journal of Cancer* 16 (1932): 1230–37.

67. Edwin B. Wilson and Ruth R. Puffer, "Least Squares and Laws of Population Growth," *Proceedings of the American Academy of Arts and Sciences* 68 (1933): 285–382.

68. E. Krummeich, "Contribution à l'étude du mouvement de la population," *Journal de la Société Statistique de Paris* 68 (1927): 119–31; 157–75; 191–99; 230–40.

69. A. T. Monk and H. R. Jeter, "The Logistic Curve and the Prediction of the Population of the Chicago Region," *Journal of the American Statistical Association* 23 (1928): 361–85. Reply by Pearl and Reed, ibid. 24 (1929): 66–67.

70. James Gray, "The Kinetics of Growth," *Journal of Experimental Biology* (British) 6 (1928–29): 248–74.

71. Hogben, "Some Biological Aspects of the Population Problem," 163–80.

72. Sewall Wright, review of *The Biology of Population Growth,* by Raymond Pearl, *Journal of the American Statistical Association* 21 (1926): 493–97. At the same time that Pearl's use of the curve was being criticized, T. B. Robertson's autocatalytic curve was coming under similar, though less polemical, attack.

73. Benjamin Gompertz, "On the Nature of the Function Expressive of the Law of Human Mortality, And on a New Mode of Determining the Value of Life Contingencies," *Philosophical Transactions of the Royal Society of London* 115 (1825): 513–85. Gompertz's curve described the number of persons living at age x, as a function of age.

74. Charles P. Winsor, "The Gompertz Curve as a Growth Curve," *Proceedings of the National Academy of Sciences of the U. S. A.* 18 (1932): 1–8. A discussion of other curves is in idem, "A Comparison of Certain Symmetrical Growth Curves," *Journal of the Washington Academy of Sciences* 22 (1932): 73–84. For a complete mathematical discussion of various growth curves, see L. G. M. Baas Becking, "On the Analysis of Sigmoid Curves," *Acta Biotheoretica* 8 (1948): 42–59.

75. See the lists of publications in "Report of the Director of the Institute for Biological Research," *Annual Report of the Johns Hopkins University*, 1925–1930.

76. For instance, J. W. MacArthur and W. H. T. Baillie, "Metabolic Activity

and Duration of Life. 1. Influence of Temperature on Longevity in *Daphnia magna,*" *Journal of Experimental Zoology* 53 (1929): 221–42; D. Stewart MacLagan, "The Effect of Population Density upon Rate of Reproduction with Special Reference to Insects," *Proceedings of the Royal Society of London,* ser. B, 111 (1932): 437–54. MacLagan started his research in W. J. Crozier's laboratory at Harvard, got his D. Sc. at Edinburgh, and became a lecturer at the University of Durham. He continued his population studies through the 1930s. See his "Recent Animal-Population Studies; And Their Significance in Relation to Socio-Biological Philosophy (Part 1)," *Proceedings of the Durham University Philosophical Society* 10 (1941): 312–31. Philippe L'Héritier and Georges Teissier, "Etude d' une population de Drosophiles en équilibre," *Comptes Rendus de l' Académie des Sciences de Paris* 197 (1933): 1765–67; idem, "Recherches sur la concurrence vitale: Etudes de populations mixtes de *Drosophila melanogaster* et de *Drosophila funebris,*" *Comptes Rendus de la Société Biologique* 118 (1935): 1396–98; W. C. Allee, *The Social Life of Animals,* rev. ed. (1951; reprint Boston: Beacon Press, 1958), chap. 5.

77. Chapman, *Animal Ecology,* 182–86.

78. Raymond Pearl, "On Biological Principles Affecting Populations; Human and Other," *American Naturalist* 71 (1937): 50–68.

79. Royal N. Chapman and L. Baird, "The Biotic Constants of *Tribolium confusum* Duval," *Journal of Experimental Zoology* 68 (1934): 293–305.

80. E. Cuyler Hammond, "Biological Effects of Population Density in Lower Organisms," *Quarterly Review of Biology* 13 (1938), 437.

81. Friedrich S. Bodenheimer, *A Biologist in Israel* (Jerusalem: Turin Press, 1959), 98.

82. Friedrich S. Bodenheimer, *Problems of Animal Ecology* (Oxford: Oxford University Press, 1938).

83. Friedrich S. Bodenheimer, *Animal Ecology Today*, vol. 6, *Monographiae Biologicae* (The Hague: W. Junk, 1958), 79.

84. Thomas Park, "Studies in Population Physiology: The Relation of Numbers to Initial Population Growth in the Flour Beetle *Tribolium confusum* Duval," *Ecology* 13 (1932): 172–81; idem, "Experimental Studies of Insect Populations," *American Naturalist* 71 (1937): 21–33.

85. F. W. Robertson and J. H. Sang, "The Ecological Determinants of Population Growth in a *Drosophila* culture," *Proceedings of the Royal Society of London,* ser. B, 132 (1944): 258–77; 277–91; J. H. Sang, "Population Growth in *Drosophila* Cultures," *Biological Reviews* 25 (1950): 188–219.

86. Bodenheimer, *Animal Ecology Today*, 78.

Chapter 5. Modeling Nature

1. Ronald Ross, "Some Quantitative Studies in Epidemiology," *Nature* 87 (1911): 466–67.

2. Ronald Ross, "An Application of the Theory of Probabilities to the Study of A Priori Pathometry," *Proceedings of the Royal Society of London*, ser. A, 92 (1916): 204–30. Ronald Ross and H. P. Hudson, ibid., ser. A, 93 (1917): 212–40.

3. John Brownlee, "Studies in Immunity: Theory of an Epidemic," *Proceedings*

of the Royal Society of Edinburgh 26 (1906): 484–521; idem, "The Mathematical Theory of Random Migration and Epidemic Distribution," ibid. 31 (1910): 262–89.

4. William H. Thorpe, "William Robin Thompson, 1887–1972," *Biographical Memoirs of the Royal Society of London* 19 (1973): 655–78.

5. W. R. Thompson to D'Arcy Thompson, 30 May 1922, in W. R. Thompson Papers, Archives of the Department of Agriculture, Ottawa, Canada.

6. Leland O. Howard, "A Study in Insect Parasitism," U.S. Department of Agriculture, Division of Entomology, Technical Series no. 5 (Washington, D. C.: 1897); Paul Marchal, "The Utilization of Auxiliary Entomophagous Insects in the Struggle against Insects Injurious to Agriculture," *Popular Science Monthly* 72 (1908): 352–70; 406–19.

7. W. R. Thompson, "La théorie mathématique de l'action des parasites entomophages," *Revue Générale des Sciences Pures et Appliquées* 34 (1923), 203.

8. Ibid., 204.

9. W. R. Thompson, "Théorie de l'action des parasites entomophages: les formules mathématiques du parasitisme cyclique," *Comptes Rendus de l'Académie des Sciences de Paris* 174 (1922): 1201–4; idem, "Etude mathématique de l'action des parasites entomophages: durée du cycle parasitaire et accroissement de la proportion d'hôtes parasites," ibid., 1433–35; idem, "Etudes de quelques cas simples de parasitisme cyclique chez les insectes entomophages," ibid., 1647–49; idem, "Théorie de l'action des parasites entomophages: accroissement de la proportion d'hôtes parasites dans le parasitisme cyclique," ibid., 175 (1922): 65–68.

10. Thompson, "Biological Control" (see chap. 3, n.7), 312; idem, "La théorie mathématique de l'action des parasites entomophages et le facteur du hasard," *Annales Faculté des Sciences de Marseille* 2 (1924): 69–89.

11. Alfred J. Lotka, "Quantitative Studies in Epidemiology," *Nature* 88 (1912): 497–98.

12. Alfred J. Lotka, "Contributions to the Analysis of Malaria Epidemiology" (in 5 parts; part 4 with F. R. Sharpe), *American Journal of Hygiene* (Suppl.) 3 (1923): 1–121.

13. W. O. Kermack and A. G. McKendrick, "A Contribution to the Mathematical Theory of Epidemics," *Proceedings of the Royal Society of London*, ser. A, 115 (1927): 700–721; idem, "Contributions to the Mathematical Theory of Epidemics. II. The Problem of Endemicity," ibid., 138 (1932): 55–83; idem, Part III, "Further Studies of the Problem of Endemicity," ibid., 141 (1933): 94–122; idem, Part IV, "Analysis of Experimental Epidemics of the Virus Disease Mouse Ectromelia," *Journal of Hygiene* (Cambridge) 37 (1937): 172–87.

14. Ross to Lotka, 6 October 1925, Box 4, Lotka Papers.

15. Anonymous review in *Science Progress* 20 (1925): 337–39.

16. Major Greenwood, *Epidemiology, Historical and Experimental* (Baltimore: Johns Hopkins University Press, 1932).

17. Bertalanffy, *General System Theory* (see chap. 2, n.6).

18. Lotka, *Elements*, 57–63.

19. Ibid., 83–92.

20. Ibid., 94–95.

21. E. T. Whittaker, "Vito Volterra, 1860–1940," *Obituary Notices of the Royal Society of London* 3 (1939–41): 691–729.

22. Vito Volterra, "Variazioni e fluttuazioni del numero d'individui in specie animali conviventi," *Memoria della Regia Accademia Nazionale dei Lincei*, ser. 6, 2 (1926): 31–113.

23. Vito Volterra, "Fluctuations in the Abundance of a Species Considered Mathematically," *Nature* 118 (1926): 558–60.

24. James P. Finerty, *The Population Ecology of Cycles in Small Mammals: Mathematical Theory and Biological Fact* (New Haven and London: Yale University Press, 1980), 127. See also Philip M. Morse and George E. Kimball, *Methods of Operations Research*, rev. ed. (London: Chapman and Hall, 1950).

25. Lotka, *Elements*, 361.

26. Described most clearly in Vito Volterra and Umberto D'Ancona, *Les associations biologiques au point de vue mathématique* (Paris: Hermann, 1935), chap. 6. See also Francesco M. Scudo, "Vito Volterra and Theoretical Ecology," *Theoretical Population Biology* 2 (1971): 1–23.

27. Letters by Lotka and Volterra, *Nature* 119 (1927): 12–13.

28. Lotka to Volterra, 13 December 1926, courtesy of Luisa Volterra D'Ancona, Pavia, Italy.

29. Undated manuscript, In Lotka Papers.

30. Vito Volterra, "Variazioni e fluttuazioni del numero d'individui in specie animali conviventi," *Regio Comitato Talassografico Italiano, Memoria* 131 (1927): 1–142; idem, "Une théorie mathématique de la lutte pour la vie," *Scientia* (Suppl.) 41 (1927): 33–48. An English translation of his 1926 article (see n.22) was published in *Journal du Conseil International pour l'Exploration de la Mer* 3 (1928): 3–51; a Russian translation was published in Moscow, also in 1928. For more information on Volterra, see Augusto G. Salvatore, "On the Applications of Mathematics to Certain Biological Problems Considered by Alfred J. Lotka (1880–1949) and Vito Volterra (1860–1940)" (Ph.D. diss., New York University, 1972).

31. Lotka to Pearl, 17 May 1928, A. J. Lotka file, Pearl Papers. See Alfred J. Lotka, "Families of Curves of Pursuit and Their Isochrones," *American Mathematical Monthly* 35 (1928): 421–24.

32. Alfred J. Lotka, "Contribution to the Mathematical Theory of Capture. I. Conditions for Capture," *Proceedings of the National Academy of Sciences of the U. S. A.* 18 (1932). 172–78.

33. Lotka to Pearl, 6 January 1931, A. J. Lotka file, Pearl Papers.

34. Alfred J. Lotka, "The Growth of Mixed Populations: Two Species Competing for a Common Food Supply," *Journal of the Washington Academy of Sciences* 22 (1932): 461–69.

35. Vito Volterra, *Leçons sur la théorie mathématique de la lutte pour la vie* (Paris: Gauthier-Villars, 1931).

36. Joseph Pérès, "Une application nouvelle des mathématiques à la biologie: La théorie des associations biologiques d'après les travaux de M. Vito Volterra," *Revue Générale des Sciences Pures et Appliquées* 38 (1927): 295–300; 337–41.

37. Lotka to Pearl, 17 May 1928, A. J. Lotka file, Pearl Papers.

38. Karl Friederichs, *Die Grundfragen und Gesetzmässigkeiten der land- und forstwirtschaftlichen Zoologie* (Berlin, 1930), 257–75; 291–98.

39. Lotka to Friederichs, 30 March 1931, Lotka Papers. Lotka was not as averse to priority disputes as he claimed, for he was at that time engaged in a vigorous dispute with another demographer, R. R. Kuczynski. See Paul A. Samuelson, "Resoving a Historical Confusion in Population Analysis," *Human Biology* 48 (1976): 559–80.

40. Francesco M. Scudo and James R. Ziegler, "Vladimir Aleksandrovich Kostitzin and Theoretical Ecology," *Theoretical Population Biology* 10 (1976): 395–412. Scudo and Ziegler also mention Kostitzin's close connection to the Russian geochemist Vladimir Vernadsky.

41. Vladimir A. Kostitzin, *Symbiose, parasitisme et évolution (étude mathématique)* (Paris: Hermann, 1934).

42. Lotka to Kostitzin, undated draft of a letter, probably written in 1933, Lotka Papers.

43. Alfred J. Lotka, *Théorie analytique des associations biologiques*, 2 vols. (Paris: Hermann, 1934 and 1939).

44. Volterra and D'Ancona, *Les associations biologiques.*

45. A manuscript of the revision is in the Lotka Papers.

46. Vito Volterra, "Principes de biologie mathématique," *Acta Biotheoretica* 3 (1937): 1–36.

47. My source of biographical information and the development of Nicholson's ideas is a letter of reminiscences to Frank N. Egerton, 25 May 1961. Courtesy of Prof. Frank Egerton, University of Wisconsin-Parkside.

48. Alexander J. Nicholson, "A New Theory of Mimicry in Insects," *Australian Zoologist* 5 (1927): 10–104.

49. Alexander J. Nicholson, "The Balance of Animal Populations," *Journal of Animal Ecology* 2 (1933): 132–78; Alexander J. Nicholson and Victor A. Bailey, "The Balance of Animal Populations," *Proceedings of the Zoological Society of London*, part 3 (1935): 551–98.

50. Victor A. Bailey, "The Interaction between Hosts and Parasites," *Quarterly Journal of Mathematics* 2 (1931): 68–77; idem, "On the Interaction between Several Species of Hosts and Parasites," *Proceedings of the Royal Society of London*, ser. A, 143 (1933): 75–88; idem, "Non-continuous Interaction between Hosts and Parasites," *Proceedings of the Cambridge Philosophical Society* 29 (1933): 487–91.

51. Alexander J. Nicholson, "The Role of Competition in Determining Animal Populations," *Journal of the Council of Scientific and Industrial Research* (Australia) 10 (1937): 101–6.

52. Alexander J. Nicholson, "Competition for Food amongst *Lucilia cuprina* larvae," *Proceedings of Eighth International Congress of Entomology* (Stockholm, 1948), 277–81.

53. Alexander J. Nicholson, "An Outline of the Dynamics of Animal Populations," *Australian Journal of Zoology* 2 (1954): 9–65.

54. Nicholson, "Balance of Animal Populations." Actually the stability relations of Nicholson and Bailey's finite-difference equations are the same as the Lotka-Volterra equations written in difference form. At the time, however, the relationship between these models was not well understood. For a recent analysis,

see Robert M. May, "On Relationships among Various Types of Population Models," *American Naturalist* 107 (1973): 46–57.

55. Bailey to Lotka, 6 February 1933, A. J. Lotka file, Pearl Papers.

56. Lotka to Bailey, 15 March 1933; Lotka to Pearl, 19 April 1933, A. J. Lotka file, Pearl Papers.

57. Pearl to Lotka, 2 May 1933, A. J. Lotka file, Pearl Papers.

58. Lotka to Bailey, 3 May 1933, A. J. Lotka file, Pearl Papers.

59. For my discussion of these points of view, I have drawn upon the categories used by Richard Levins, "The Strategy of Model Building in Population Biology," *American Scientist* 54 (1966): 421–31.

60. Vito Volterra, "Il Momento Scientifico Presente e la Nuova Societa Italiana per il Progresso delle Scienze," in vol. 3 of *Opere matematiche* (Rome: Accademia Nazionale dei Lincei, 1962), 248. Translated and quoted by Salvatore, "On the Applications of Mathematics to Certain Biological Problems," 109–10.

61. Salvatore, "On the Applications of Mathematics to Certain Biological Problems," 107 n. 59.

62. See J. H. Poynting's address to the British Association for the Advancement of Science, meeting in Dover, 1899, in which he expresses his cautious attitude toward analogy. *Report of the British Association for the Advancement of Science* (London: John Murray, 1900), 615–24.

63. Lotka, "Law of Evolution" (see chap. 2, n.37), 167–94.

64. Ibid., 179.

65. Ibid., 172–73.

66. Lotka to Volterra, 1 December 1926, courtesy of Luisa Volterra D'Ancona.

67. Francis Bacon, *Novum Organum*, in The World's Greatest Literature series, rev. ed. (New York: P. F. Collier, 1900), preface.

Chapter 6. Skeptics and Converts

1. Harry S. Smith, "The Fundamental Importance of Life-History Data in Biological Control Work," *Journal of Economic Entomology* 19 (1926): 708–14; Smith to R. N. Chapman, 30 April 1929, no. 496, ALD9E.1, Division of Entomology and Economic Zoology Papers, 1890–1959, University of Minnesota Archives, Minneapolis.

2. William R. Thompson, *The Biological Control of Insect and Plant Pests: A Report on the Organisation and Progress of the Work of Farnham House Laboratory* (London: King's Printer, 1930), 51–56; idem, "The Utility of Mathematical Methods in Relation to Work on Biological Control," *Annals of Applied Biology* 17 (1930): 641–48.

3. W. C. Cook to Lotka, 23 February 1927, Box 4, Lotka Papers.

4. Royal N. Chapman, "The Potentialities of Entomology," *Science* 69 (1929): 413–18.

5. Royal N. Chapman, "The Quantitative Analysis of Environmental Factors," *Ecology* 9 (1928): 111–22.

6. Chapman, *Animal Ecology*.

7. These were: Robert Wardle from England (1927–1928); Karl Friederichs from Germany (1928–1929); Filippo Silvestri from Italy (1929–1930); and Friedrich S. Bodenheimer from Jerusalem (1931).

8. Royal N. Chapman, "The Measurement of the Effects of Ecological Factors," *International Congress of Entomology, Transactions* (Ithaca, 1928), vol. 2, 408–11. Forbes's comments, 411.

9. Charles Elton, review in *Journal of Animal Ecology* 4 (1935): 148–49.

10. Elton, *Animal Ecology*, 2d ed. (New York: Macmillan, 1935). See notes added to text.

11. Royal N. Chapman, "Insect Population Problems in Relation to Insect Outbreak," *Ecological Monographs* 9 (1939): 261–69.

12. Eric Ponder, remarks following Edwin B. Wilson, "Mathematics of Growth," *Cold Spring Harbor Symposia on Quantitative Biology* 2 (1934), 201.

13. Charles P. Winsor, "Mathematical Analysis of Growth of Mixed Populations," *Cold Spring Harbor Symposia on Quantitative Biology* 2 (1934): 181.

14. Royal N. Chapman, "The Causes of Fluctuations of Populations in Insects," *Proceedings of the Hawaiian Entomological Society* 8 (1933): 279–92.

15. Charles Elton, review of article by R. N. Chapman, *Journal of Animal Ecology* 4 (1935): 295–96.

16. Egon S. Pearson, "The Application of the Theory of Differential Equations to the Solution of Problems Connected with the Interdependence of Species," *Biometrika* 19 (1927): 216–22; 242.

17. Friedrich S. Bodenheimer, "Uber den Massenwechsel der Selachierbevölkerung im oberadriatischen Benthos," *Archiv für Hydrobiologie* 24 (1932): 667–75.

18. Volterra and D'Ancona, *Les associations biologiques*, 75–78.

19. S. A. Severtzov, "On the Dynamics of Populations of Vertebrates," *Quarterly Review of Biology* 9 (1934): 409–37.

20. John Stanley, "A Mathematical Theory of the Growth of Populations of the Flour Beetle, *Tribolium confusum* Duval," *Canadian Journal of Research* 6 (1932): 632–71; 7 (1932): 426–33; 450; F. G. Holdaway, "An Experimental Study of the Growth of Populations of the 'Flour Beetle,' *Tribolium confusum* Duval, As Affected by Atmospheric Moisture," *Ecological Monographs* 2 (1932): 261–304; N. M. Payne, "The Differential Effect of Environmental Factors upon *Microbracon hebetor* Say (Hymenoptera: Braconidae) and its Host, *Ephestia kühniella* Zeller (Lepidoptera: Pyralidae)," *Ecological Monographs* 4 (1934): 1–46.

21. George Salt, "Experimental Studies in Insect Parasitism. I. Introduction and Technique," *Proceedings of the Royal Society of London*, ser. B, 114 (1934): 450–54; idem, Part II, "Superparasitism," ibid. 114 (1934): 455–76.

22. L'Héritier and Teissier, "Etude d'une population de Drosophiles" (see chap. 4, n. 76), 1765–67; idem, "Recherches sur la concurrence vitale" (see chap. 4, n. 76), 1396–98; Bodenheimer, *Problems of Animal Ecology*.

23. Alfred J. Lotka, "The Stability of the Normal Age Distribution," *Proceedings of the National Academy of Sciences of the U.S.A.* 8 (1922): 339–45; Dublin and Lotka, "On the True Rate of Natural Increase" (see chap. 4, n.35), 305–39.

24. Patrick H. Leslie and R. M. Ranson, "The Mortality, Fertility and Rate of Natural Increase of the Vole (*Microtus agrestis*) As Observed in the Laboratory," *Journal of Animal Ecology* 9 (1940): 27–52.

25. Patrick H. Leslie, "On the Use of Matrices in Certain Population Mathematics," *Biometrika* 33 (1945): 183–212; idem, "Some Further Notes on the Use of Matrices in Population Mathematics," *Biometrika* 35 (1945): 213–45.

26. Charles Elton to Lotka, 11 March 1948, Lotka Papers.

27. Charles Elton, *Voles, Mice and Lemmings: Problems in Population Dynamics* (Oxford: Clarendon, 1942), 110.

28. P. H. Leslie, review of *La Lotta per l'Esistenza*, by U. D'Ancona, *Journal of Animal Ecology* 15 (1946), 107.

29. G. Currie and J. Graham, "Growth of Scientific Research in Australia: The Council for Scientific and Industrial Research and the Empire Marketing Board," *Australian Academy of Science Records* 1 (1968): 25–35.

30. William R. Thompson, "The Principles of Biological Control," *Annals of Applied Biology* 17 (1930): 306–38.

31. Thorpe, "William Robin Thompson," 661.

32. William R. Thompson, "Research on Biological Control," *Nature* 139 (1937): 552.

33. William R. Thompson, *Science and Common Sense: An Aristotelian Excursion* (London: Longmans, Green, 1937).

34. Thorpe, "William Robin Thompson," 661.

35. Fisher, *Genetical Theory of Natural Selection*, ix.

36. Thompson, *Science and Common Sense*, 119.

37. Thompson, review of *Evolution in the Genus Drosophila*, by J. T. Patterson and W. S. Stone, *Canadian Entomologist* 86 (1954): 98–100.

38. Thompson, *Science and Common Sense*, 231.

39. Volterra, "Principes de biologie mathématique," 1036.

40. Thompson, "Biological Control" (see chap. 3, n.7), 299–388.

41. William R. Thompson, "Can Economic Entomology Be an Exact Science?" *Canadian Entomologist* 80 (1948): 49–55.

42. Thompson, *Science and Common Sense*, 218–19.

43. Thompson, "Biological Control," 337–38.

44. Ernst Mayr, "Where Are We?" *Genetics and Twentieth Century Darwinism. Cold Spring Harbor Symposia on Quantitative Biology* 24 (1959), 6.

45. Harry S. Smith, "Insect Populations in Relation to Biological Control," *Ecological Monographs* 9 (1939): 311–20. See also idem, "The Role of Biotic Factors in the Determination of Population Densities," *Journal of Economic Entomology* 28 (1935): 873–98.

46. Nicholson, "An Outline of the Dynamics of Animal Populations," 54.

47. Nicholson to Frank N. Egerton, 25 May 1961, courtesy of Prof. Egerton.

48. Paul DeBach and H. S. Smith, "Are Population Oscillations Inherent in Nature?" *Ecology* 22 (1941): 363–69.

49. J. B. S. Haldane, *The Causes of Evolution* (New York and London: Harper & Brothers, 1932), 214.

50. Ernst Mayr, "Some Thoughts on the History of the Evolutionary Synthe-

sis," in *The Evolutionary Synthesis: Perspectives on the Unification of Biology*, ed. Ernst Mayr and W. B. Provine (Cambridge, Mass.: Harvard University Press, 1980), 1–48.

51. These were originally published from 1936 to 1939 in *Comptes Rendus de l'Académie des Sciences de Paris*. Five of Kostitzin's articles have been translated into English and reprinted in F. M. Scudo and J. R. Ziegler, eds., *The Golden Age of Theoretical Ecology, 1923–1940* (Berlin: Springer-Verlag, 1978).

52. Scudo and Ziegler, "Vladimir Aleksandrovich Kostitzin," 395–412.

53. E. B. Ford, "Some Recollections Pertaining to the Evolutionary Synthesis," in *The Evolutionary Synthesis*, ed. Mayr and Provine, 334–42.

54. Charles Elton, "Animal Numbers and Adaptation," in *Evolution: Essays on Aspects of Evolutionary Biology Presented to Professor E. S. Goodrich*, ed. Gavin R. de Beer (Oxford: Clarendon, 1938), 127–37.

55. Thomas Park, "Integration in Infra-social Insect Populations," *Biological Symposia* 8 (1942), 122.

Chapter 7. The Niche, the Community, and Evolution

1. Professor of histology at the University of Moscow from 1924 to 1929; author of the theory of "mitogenetic rays," weak shortwave ultraviolet radiation thought to be a causal factor in mitosis.

2. Alpatov to Pearl, 1 August 1926, W. W. [V. V.] Alpatov file, Pearl Papers.

3. Alpatov to Pearl, 15 September 1929, Alpatov file, Pearl Papers.

4. V. Alpatov, report to Dr. Tisdale (Paris Office, Rockefeller Foundation), undated but probably written in August 1929, Alpatov file, Pearl Papers.

5. Alpatov to Pearl, 9 January 1930, Alpatov file, Pearl Papers.

6. Alpatov to Pearl, 12 April 1931, Alpatov file, Pearl Papers.

7. Alpatov to Pearl, 15 October 1932, Alpatov file, Pearl Papers.

8. Georgii F. Gause, "Studies on the Ecology of the Orthoptera," *Ecology* 11 (1930): 307–25.

9. Georgii F. Gause, "The Influence of Ecological Factors on the Size of Population," *American Naturalist* 65 (1931): 70–76. Data were from Arata Terao and Royal N. Chapman.

10. Georgii F. Gause, "Ecology of Populations," *Quarterly Review of Biology* 7 (1932): 27–48.

11. G. F. Gause and W. W. [V. V.] Alpatov, "Die logistische Kurve von Verhulst-Pearl und ihre Anwendung im Gebiet der quantitativen Biologie," *Biologisches Zentralblatt* 51 (1931): 1–14.

12. Pearl informed Alpatov that he had sent such a letter, along with Gause's articles, to the Rockefeller Foundation, in a letter dated 29 April 1931, but I have not seen the letter itself. Alpatov file, Pearl Papers.

13. Alpatov to Pearl, 3 June 1932, Alpatov file, Pearl Papers.

14. Letter to Alpatov from someone in Pearl's office, informing him of the news about Gause, 16 July 1932, Alpatov file, Pearl Papers.

15. Pearl to Alpatov, 15 December 1932, Alpatov file, Pearl Papers.

16. Oscar W. Richards, "The Growth of the Yeast *Saccharomyces cerevisiae*. I.

The Growth Curve, Its Mathematical Analysis, and the Effect of Temperature on the Yeast Growth," *Annals of Botany* 42 (1928): 271–83; "The Second Cycle and Subsequent Growth of a Population of Yeast," *Archiv für Protistenkunde* 78 (1932): 263–301.

17. For information on the history of Russian ecology, I am indebted to Douglas R. Weiner, "The History of the Conservation Movement in Russia and the U.S.S.R. from Its Origins to the Stalin Period" (Ph.D. diss., Columbia University, 1983).

18. Translated in French as *La Biosphère.*

19. Georgii F. Gause, *The Struggle for Existence* (1934; reprint New York: Dover, 1971), 114–40.

20. Ibid., 133–35.

21. Georgii F. Gause, *Vérifications expérimentales de la théorie mathématique de la lutte pour la vie* (Paris: Hermann, 1935), 61.

22. Gause, *Struggle for Existence*, 43.

23. Ibid., 35. I have changed the notation for consistency.

24. Hutchinson, *Introduction to Population Ecology* (see chap. 3, n.45), 23.

25. Lotka, "Growth of Mixed Populations, 461–69.

26. Lotka gave no reference for this criticism, but he was probably referring to a review by Joseph Bertrand, "Théorie des richesses," *Journal des Savants* (1883): 499–508. See also Robert H. MacArthur, "Species Packing and Competitive Equilibrium for Many Species," *Theoretical Population Biology* 1 (1970): 1–11.

27. A discussion of this point was later published by Levins, "Strategy of Model Building, 421–31.

28. Gause, *Struggle for Existence*, 58.

29. John L. Harper, "A Centenary in Population Biology," *Nature* 252 (1974): 526–27.

30. Hutchinson, *Introduction to Population Ecology*, 152–57. See also David L. Cox, "A Note on the Queer History of the Niche," *Bulletin of Entomological Society of America* 61 (1980): 201–2.

31. Joseph Grinnell, "The Niche-Relationships of the California Thrasher," *Auk* 34 (1917): 427–33.

32. J. B. S. Haldane, "A Mathematical Theory of Natural and Artificial Selection," *Transactions of the Cambridge Philosophical Society* 23 (1924): 19–41.

33. Gause, *Struggle for Existence*, 48, 98.

34. Elton, *Animal Ecology*, 63–64.

35. Charles Elton, *The Ecology of Animals* (London: Methuen, 1933), 28.

36. Georgii F. Gause, "The Principles of Biocoenology," *Quarterly Review of Biology* 11 (1936): 320–36; idem, "Experimental Populations of Microscopic Organisms," *Ecology* 18 (1937): 173–79.

37. Georgii F. Gause, remarks following a paper by Thomas Park, "Analytical Population Studies in Relation to General Ecology," *American Midland Naturalist* 21 (1939): 235–53. Quotation on p. 255.

38. Garrett Hardin, "The Competitive Exclusion Principle," *Science* 131 (1960): 1292–97.

39. Gause, "Experimental Populations," 173–79.

40. Alpatov to Pearl, 24 June 1940, Alpatov file, Pearl Papers.

41. N. V. Minin, "Onekotorykh idealisticheskikh ucheniiakh v ekologii," *Priroda*, no. 7 (1939): 30–43. This article was brought to my attention by Douglas Weiner. My discussion is drawn from a paper presented by Mr. Weiner, "Socialist Science, Bourgeois Science, and Ecology: The Untimely Demise of Community Ecology in Stalin's Russia," at the Joint Atlantic Seminar in the History of Biology, April 1983, John Hopkins University.

42. Georgii F. Gause, "The Effect of Natural Selection in the Acclimatization of *Euplotes* to Different Salinities of the Medium," *Journal of Experimental Zoology* 87 (1941): 85–100; G. F. Gause, N. P. Smaragdova, and W. W. [V. V.] Alpatov, "Geographic Variation in *Paramecium* and the Role of Stabilizing Selection in the Origin of Geographic Differences," *American Naturalist* 76 (1942): 63–74; G. F. Gause and W. W. [V. V.] Alpatov, "On the Inverse Relation between Inherent and Acquired Properties of Organisms," *American Naturalist* 79 (1945): 478–80.

43. George Gaylord Simpson, "The Baldwin Effect," *Evolution* 7 (1953): 110–17; C. H. Waddington, "Genetic Assimilation of an Acquired Character," *Evolution* 7 (1953): 118–26.

44. Georgii F. Gause, "Problems of Evolution," *Transactions of the Connecticut Academy of Arts and Sciences* 37 (1947): 17–68.

45. Georgii F. Gause, *Optical Activity and Living Matter* (1941), reprinted from *Biodynamica*, nos. 52 and 56 (1939); nos. 62 and 63 (1940); nos. 70 and 71 (1941).

46. See the report of the conference, *The Situation in Biological Science*, Verbatim Report, Lenin Academy of Agricultural Sciences of the U.S.S.R. (Moscow: 1949).

47. "Why Russia Purges Scientists," *U.S. News & World Report* 26 (14 January 1949): 22–23.

48. R. C. Cook, "Lysenko's Wonderful Genetics," *Journal of Heredity* 40 (1949): 169–208; Julian S. Huxley, *Heredity East and West: Lysenko and World Science* (New York: Henry Schuman, 1949), 193–94 n.

49. For a general history of Lysenkoism, see David Joravsky, *The Lysenko Affair* (Cambridge, Mass.: Harvard University Press, 1970).

50. Elton, *Voles, Mice, and Lemmings*. Charles Elton, review of *The Struggle for Existence* by G. F. Gause, *Journal of Animal Ecology* 4 (1935): 294–95; Royal N. Chapman, review in *Ecology* 16 (1935): 656–57; Thomas Park, review in *Quarterly Review of Biology* 10 (1935): 209–12; Park, "Analytical Population Studies," 235–53.

51. David L. Lack, *Darwin's Finches* (Cambridge: Cambridge University Press, 1947), 11.

52. Harry S. Swarth, "The Bird Fauna of the Galapagos Islands in Relation to Species Formation," *Biological Reviews* 9 (1934): 213–34.

53. Malcolm J. Kottler, "Isolation and Speciation, 1837–1900" (Ph.D. diss., Yale University, 1976), chap. 4.

54. Nicholson, "A New Theory of Mimicry (see chap. 5, n.48), 10–104.

55. O. W. Richards and G. C. Robson, "The Species Problem and Evolution," *Nature* 117 (1926): 345–47; 382–84. See also Robson and Richards, *Variation of Animals in Nature* (see chap. 3, n.14), 352–55.

56. Marjorie Grene, ed., *Dimensions of Darwinism* (Cambridge: Cambridge University Press, 1983).

57. Elton, *The Ecology of Animals.*

58. Haldane, *The Causes of Evolution*, 201–4; Sewall Wright, "Breeding Structure of Populations in Relation to Speciation," *American Naturalist* 74 (1940): 232–48.

59. E. B. Ford, *Mendelism and Evolution* (London: Methuen, 1931).

60. H. D. Ford and E. B. Ford, "Fluctuation in Numbers and Its Influence on Variation in *Melitaea aurinia*," *Transactions of Royal Entomological Society of London* 78 (1930): 345–51. E. B. Ford was very influenced by R. A. Fisher, "On Some Objections to Mimicry Theory: Statistical and Genetic," ibid., 75 (1927): 269–78. See his comments on the period of the synthesis in "Some Recollections," in *The Evolutionary Synthesis*, ed. Mayr and Provine, 334–42.

61. *Proceedings of the Linnean Society of London*, 147th session (1935): 87.

62. Julian S. Huxley, "Natural Selection and Evolutionary Progress," *Report of the British Association for the Advancement of Science* (Blackpool, 1936), 87.

63. See, for instance, "A Discussion on the Present State of the Theory of Natural Selection," *Proceedings of the Royal Society of London*, ser. B, 121 (1936): 43–73.

64. Percy R. Lowe, "The Finches of the Galapagos in Relation to Darwin's Conception of Species," *Ibis* 6 (1936): 310–21.

65. David Lack, "My Life As an Amateur Ornithologist," *Ibis* 115 (1973): 421–31.

66. Lack, *Darwin's Finches*, 1.

67. Lack, "My Life." See also Mayr's comments following, pp. 432–34.

68. David L. Lack, "The Galapagos Finches (*Geospizinae*); A Study in Variation," *Occasional Papers of the California Academy of Sciences*, no. 21 (San Francisco, 1945).

69. G. E. Hutchinson, personal communication, from his correspondence with Formosov's son.

70. Julian S. Huxley, *Evolution: The Modern Synthesis*, American ed. (New York: Harper & Brothers, 1943), 154–56.

71. David Lack, "Habitat Selection in Birds with Special Reference to the Effects of Afforestation on the Breckland Avifauna," *Journal of Animal Ecology* 2 (1933): 239–62.

72. J. S. Huxley, review of *Systematics and the Origin of Species*, by Ernst Mayr, *Nature* 151 (1943): 347–48.

73. George C. Varley, "The Natural Control of Population Balance in the Knapweed Gall-fly (*Urophora jaceana*)," *Journal of Animal Ecology* 16 (1947): 139–86.

74. George Varley, personal communication.

75. Lack, *Darwin's Finches*, 136.

76. Lack, "My Life," 429.

77. David Lack, "Competition for Food by Birds of Prey," *Journal of Animal Ecology* 15 (1946): 123–29.

78. David Lack, "Ecological Aspects of Species-Formation in Passerine Birds," *Ibis* 86 (1944): 260–86.

79. "Symposium on 'The Ecology of Closely Allied Species,'" *Journal of*

Animal Ecology 13 (1944): 176–77. See also David Lack, "The Ecology of Closely Related Species with Special Reference to Cormorant (*Phalacrocorax carbo*) and Shag (*P. aristotelis*)," *Journal of Animal Ecology* 14 (1945): 12–16. Elton's argument appeared as "Competition and the Structure of Ecological Communities," ibid. 15 (1946): 54–68.

80. Frank J. Sulloway, "Darwin and His Finches: The Evolution of a Legend," *Journal of the History of Biology* 15 (1982): 1–53.

81. Roger Lewin, "Finches Show Competition in Ecology," *Science* 219 (1983): 1411–12. See also the introductory notes and bibliography by L. M. Ratcliffe and P. T. Boag to the recent edition of Lack, *Darwin's Finches* (Cambridge: Cambridge University Press, 1983). These compare Lack's ideas to modern studies.

82. See, for instance, O. Gilbert, T. B. Reynoldson, and J. Hobart, "Gause's Hypothesis: An Examination," *Journal of Animal Ecology* 21 (1952): 310–12; Hardin, "The Competitive Exclusion Principle," *Science* 131 (1960): 1292–97, and ensuing debates in *Science*, vols. 132, 133, 134; Lawrence B. Slobodkin, *Growth and Regulation of Animal Populations* (New York: Holt, Rinehart and Winston, 1961); Paul DeBach, "The Competitive Displacement and Coexistence Principles," *Annual Review of Entomology* 11 (1966): 183–212.

83. Thomas Park, "Some Observations on the History and Scope of Population Ecology," *Ecological Monographs* 16 (1946): 315–20.

84. Thomas Park, "Experimental Studies of Interspecies Competition. II. Temperature, Humidity and Competition in Two Species of *Tribolium*," *Physiological Zoology* 27 (1954): 177–238. The stages of Park's program are summarized in T. Park, P. H. Leslie, and D. B. Mertz, "Genetic Strains and Competition in Populations of Tribolium," ibid. 37 (1964): 97–162.

85. Lack's review of some of these controversies is in his appendix to *Population Studies of Birds* (Oxford: Clarendon, 1966), 281–312. There is an odd contrast between Lack's personal style and his scientific style. In person he was extremely shy and modest, though when the shyness evaporated he was very warm. He was also a man of deep moral convictions and converted to Christianity in 1948, a conversion he attributed to the influence of his Dartington friends. His writings had the straightforward, logical clarity of a didactic exercise, probably the product of his years of teaching experience. Mayr recalls that Lack's enthusiasm and dedication would lead him to adopt single-factor explanations, a tendency which no doubt contributed to the frequent, intense controversies in which he found himself. But for Lack, scientific disagreements had no personal overtones. Mayr remembers their own arguments having no effect on their intimate friendship. See Lack, "My Life," and Mayr's comments following.

86. David Lack, *The Natural Regulation of Animal Numbers* (Oxford: Clarendon, 1954).

87. H. G. Andrewartha and L. C. Birch, *The Distribution and Abundance of Animals* (Chicago and London: University of Chicago Press, 1954).

88. John R. Baker, "The Evolution of Breeding Seasons," in *Evolution: Essays on Aspects of Evolutionary Biology*, ed. de Beer, 161–77.

89. L. Charles Birch, "Experimental Background to the Study of the Distribution and Abundance of Insects. III. The Relation between Innate Capacity for

Increase and Survival of Different Species of Beetles Living Together on the Same Food," *Evolution* 7 (1953): 136–44.

90. Alistair C. Crombie, "Interspecific Competition," *Journal of Animal Ecology* 16 (1947): 44–73.

91. C. B. Huffaker, "Experimental Studies on Predation: Dispersion Factors and Predator-Prey Oscillations," *Hilgardia* 27 (1958): 343–83; Nicholson, "Outline of the Dynamics of Animal Populations," 9–65; George C. Varley, "Ecology as an Experimental Science," *Journal of Ecology* 45 (1957): 639–48; M. E. Solomon, "The Natural Control of Animal Populations," *Journal of Animal Ecology* 18 (1949): 1–35.

92. J. G. Skellam, "The Mathematical Approach to Population Dynamics," in *The Numbers of Man and Animals*, ed. J. B. Cragg and N. W. Pirie (Edinburgh: Oliver and Boyd, 1955); David B. Mertz, "The *Tribolium* Model and the Mathematics of Population Growth," *Annual Review of Ecology and Systematics* 3 (1972): 51–78.

93. Maurice S. Bartlett, *Stochastic Population Models in Ecology and Epidemiology* (London: Methuen, 1960).

94. V. A. Bailey, A. J. Nicholson, and E. J. Williams, "Interaction between Hosts and Parasites When Some Host Individuals Are More Difficult To Find Than Others," *Journal of Theoretical Biology* 3 (1962): 1–18. For a more recent analysis of the different models, see May, "On Relationships among Various Types of Population Models," 46–57.

95. Frederick E. Smith, "Experimental Methods in Population Dynamics: A Critique," *Ecology* 33 (1952): 441–50.

96. "Population Studies: Animal Ecology and Demography," *Cold Spring Harbor Symposia on Quantitative Biology* 22 (1957).

97. Lamont C. Cole, "Population Cycles and Random Oscillations," *Journal of Wildlife Management* 15 (1951): 233–52; idem, "Some Features of Random Population Cycles," ibid. 18 (1954): 1–24. The analytical techniques developed to determine the reality of these cycles are discussed in Finerty, *The Population Ecology of Cycles in Small Mammals*.

98. P. A. P. Moran, "The Statistical Analysis of the Canadian Lynx Cycle," *Australian Journal of Zoology* 1 (1953): 163–73; 291–98.

99. Comment by T. Dobzhansky in "Population Studies: Animal Ecology and Demography," *Cold Spring Harbor Symposia*, vol. 22 (1957): 235.

100. Quoted by A. J. Nicholson, "The Self-adjustment of Populations to Change," *Cold Spring Harbor Symposia*, vol. 22 (1957): 154.

101. Ibid., 153.

102. The first comprehensive ecology text to give adequate attention to evolution was W. C. Allee *et al.*, *Principles of Animal Ecology* (Philadelphia and London: W. B. Saunders, 1949). Unfortunately, the section on evolution, written by A. E. Emerson, did not take into account the advances of the previous decade.

103. Lamont C. Cole, "The Population Consequences of Life-History Phenomena," *Quarterly Review of Biology* 29 (1954): 103–37.

104. Ernst Mayr, "Cause and Effect in Biology," *Science* 134 (1961): 1501–6.

105. G. H. Orians, "Natural Selection and Ecological Theory," *American Naturalist* 96 (1962): 257–63. See also John L. Harper, "Approaches to the Study

of Plant Competition," in *Mechanisms in Biological Competition*, Symposium in Experimental Biology, no. 15 (Cambridge: Cambridge University Press, 1961), 1–39.

Chapter 8. The Eclipse of History

1. Robert H. MacArthur, review of *Cold Spring Harbor Symposia*, "Population Studies: Animal Ecology and Demography," *Quarterly Review of Biology* 35 (1960): 82–83.

2. G. E. Hutchinson, "Concluding Remarks," *Cold Spring Harbor Symposia on Quantitative Biology* 22 (1957): 415–27.

3. G. E. Hutchinson, *The Kindly Fruits of the Earth: Recollections of an Embryo Ecologist* (New Haven: Yale University Press, 1979).

4. Goldschmidt was an important geochemist, who was also a friend of Hutchinson's father. His ideas were applied to biogeochemical problems in G. E. Hutchinson, "The Biogeochemistry of Aluminum and of Certain Related Elements," *Quarterly Review of Biology* 18 (1943): 1–29; 128–53; 242–62; 331; 363. The work on aluminum largely arose from Vernadsky's belief that the elementary composition of an organism might be a specific character. Hutchinson, personal communication, 1983.

5. G. E. Hutchinson, "Ecological Aspects of Succession in Natural Populations," *American Naturalist* 75 (1941): 406–18.

6. G. E. Hutchinson, "The Paradox of the Plankton," *American Naturalist* 95 (1961): 137–45; idem, *The Ecological Theater and the Evolutionary Play* (New Haven and London: Yale University Press, 1965); idem, *Introduction to Population Ecology*, 237–39.

7. Harold Jeffreys, *Scientific Inference* (Cambridge: Cambridge University Press, 1931), vii.

8. G. E. Hutchinson, personal communication, 12 January 1984.

9. Jeffreys, *Scientific Inference*, 6.

10. G. E. Hutchinson and E. S. Deevey, Jr., "Ecological Studies on Populations," *Survey of Biological Progress*, (New York: Academic Press, 1949), vol. 1, 325–59; G. E. Hutchinson, "The Concept of Pattern in Ecology," *Proceedings of the Academy of Natural Sciences of Philadelphia* 105 (1953): 1–12.

11. G. E. Hutchinson, review of *Voles, Mice, and Lemmings*, by Charles Elton, *Quarterly Review of Biology* 17 (1942): 354–57.

12. G. E. Hutchinson, "Circular Causal Systems in Ecology," *Annals of the New York Academy of Sciences* 50 (1948): 221–46.

13. G. E. Hutchinson, "A Note on the Theory of Competition between Two Social Species," *Ecology* 28 (1947): 319–21.

14. Slobodkin, *Growth and Regulation of Animal Populations*.

15. For instance, Edward S. Deevey, "Life Tables for Natural Populations of Animals," *Quarterly Review of Biology* 22 (1947): 283–314; Lawrence B. Slobodkin, "Population Dynamics in *Daphnia obtusa* Kurz," *Ecological Monographs* 24 (1954): 69–88; F. E. Smith, "Experimental Methods in Population Dynamics, 441–50.

16. Gordon A. Riley, introduction to *Limnology and Oceanography* 16 (1971): 177.

17. J. W. MacArthur and Baillie, "Metabolic Activity and Duration of Life. 1" (see chap. 4, n.76), 221–42.

18. It was not Hutchinson alone, but the students he had in the 1950s who contributed to the exciting intellectual atmosphere of his advanced ecology seminar and who helped to produce many of the ideas which finally went into the "Concluding Remarks." Hutchinson acknowledged his indebtedness to this stimulating group of people, which, in addition to MacArthur, included Jane Brower, Lincoln Brower, Joseph Frankel, Alan Kohn, Peter Klopfer, Gordon A. Riley, Peter Wangersky, and Sally Wheatland. (The remaining name mentioned, J. C. Foothills, was a fictional character inserted by the editor of the symposium.)

19. Martin L. Cody and Jared M. Diamond, eds., *Ecology and Evolution of Communities* (Cambridge, Mass., and London: Belknap Press, 1975), vii.

20. Edward Haskell, "Mathematical Systematization of 'Environment,' 'Organism,' and 'Habitat,'" *Ecology* 21 (1940): 1–16.

21. V. A. Kostitzin, "Evolution of the Atmosphere, Organic Circulation, Glacial Periods," in *The Golden Age of Theoretical Ecology, 1923–1940*, ed. Scudo and Ziegler, 439–84; Hutchinson, *Introduction to Population Ecology*, 158 n. 25.

22. Hutchinson, "Concluding Remarks," 416.

23. Hutchinson, *Introduction to Population Ecology*, 158–59.

24. Robert H. MacArthur, "Population Ecology of Some Warblers of Northeastern Coniferous Forests," *Ecology* 39 (1958): 599–619.

25. This is a point stressed by those who knew MacArthur and one not always appreciated by his critics. MacArthur learned much of his mathematics from William Feller's *Introduction to Probability Theory and Its Applications*, first published in 1948. Feller, who taught at Princeton, had been interested in the logistic curve, Volterra's models, and Lotka's demographic work as early as the 1930s, though his text does not emphasize these applications.

26. Robert H. MacArthur, *Geographical Ecology* (New York: Harper & Row, 1972), 169.

27. A brief review is in Nelson G. Hairston, "Species Abundance and Community Organization," *Ecology* 40 (1959): 404–16.

28. Elton, *Animal Ecology*.

29. R. A. Fisher, *Statistical Methods for Research Workers* (Edinburgh: Oliver and Boyd, 1925).

30. R. A. Fisher, A. S. Corbet, and C. B. Williams, "The Relation between the Number of Species and the Number of Individuals in a Random Sample of an Animal Population," *Journal of Animal Ecology* 12 (1943): 42–58.

31. Frank W. Preston, "The Commonness, and Rarity, of Species," *Ecology* 29 (1948): 254–58.

32. Hutchinson, "The Concept of Pattern in Ecology," 1–12.

33. D. E. Barton and F. N. David, "Some Notes on Ordered Random Intervals," *Journal of the Royal Statistical Society of London*, ser. B, 18 (1956): 79–94.

34. Robert H. MacArthur, "On the Relative Abundance of Bird Species," *Proceedings of the National Academy of Sciences of the U. S. A.* 43 (1957):

293–95; idem, "On the Relative Abundance of Species," *American Naturalist* 94 (1960): 25–36.

35. G. E. Hutchinson, "Homage to Santa Rosalia, or Why Are There So Many Kinds of Animals?" *American Naturalist* 93 (1959): 145–59.

36. G. E. Hutchinson, personal communication, 12 January 1984.

37. W. L. Brown and E. O. Wilson, "Character Displacement," *Systematic Zoology* 5 (1956): 49–64.

38. Robert H. MacArthur and Richard Levins, "Competition, Habitat Selection, and Character Displacement in a Patchy Environment," *Proceedings of the National Academy of Sciences of the U. S. A.* 51 (1964): 1207–10; idem, "The Limiting Similarity, Convergence and Divergence of Coexisting Species," *American Naturalist* 101 (1967): 377–85.

39. Robert H. MacArthur, "Patterns of Species Diversity," *Biological Reviews* 40 (1965): 510–33.

40. Robert H. MacArthur, comment following E. C. Pielou, "Comment on a Report by J. H. Vandermeer and R. H. MacArthur concerning the Broken Stick Model of Species Abundance," *Ecology* 47 (1966), 1074.

41. There is by now a large literature on the mathematical models of abundance and diversity, including much detailed analysis of the broken-stick distribution. This has been discussed by Robert M. May, "Patterns of Species Abundance and Diversity," in *Ecology and Evolution of Communities*, ed. Cody and Diamond, 81–120.

42. Levins, "The Strategy of Model Building, 421–31.

43. Ibid., 423.

44. G. E. P. Box and S. L. Andersen, "Permutation Theory in the Derivation of Robust Criteria and the Study of Departures from Assumption," *Journal of the Royal Statistical Society*, ser. B, 17 (1955): 1–26.

45. J. G. Skellam, "The Mathematical Approach to Population Dynamics," in *The Numbers of Man and Animals*, ed. Cragg and Pirie, 31–46.

46. William C. Wimsatt, "Randomness and Perceived-Randomness in Evolutionary Biology," *Synthese* 43 (1980), 307.

47. Robert H. MacArthur, "The Theory of the Niche," in *Population Biology and Evolution*, ed. Richard C. Lewontin (Syracuse, N.Y.: Syracuse University Press, 1966), 159.

48. MacArthur, review in *Quarterly Review of Biology* 35 (1960), 82.

49. MacArthur, "Theory of the Niche," 162.

50. Robert H. MacArthur, "Some Generalized Theorems of Natural Selection," *Proceedings of the National Academy of Sciences of the U. S. A.* 48 (1962): 1893–97; idem, "Ecological Consequences of Natural Selection," in *Theoretical and Mathematical Biology*, ed. Talbot H. Waterman and Harold J. Morowitz (New York: Blaisdell, 1965), 388–97.

51. MacArthur, "Patterns of Species Diversity," 510.

52. Robert H. MacArthur, "Population Effects of Natural Selection," *American Naturalist* 95 (1961), 195.

53. MacArthur, *Geographical Ecology*, 1.

54. Ibid., 177.

55. Robert H. MacArthur, "Patterns of Terrestrial Bird Communities," in

Avian Biology, ed. Donald S. Farmer and James R. King (New York and London: Academic Press, 1971), 219–20.

56. Ibid., 189–90.

57. MacArthur, *Geographical Ecology*, xi.

58. R. H. MacArthur and E. O. Wilson, "An Equilibrium Theory of Insular Zoogeography," *Evolution* 17 (1963): 373–87.

59. R. H. MacArthur and E. O. Wilson, *The Theory of Island Biogeography* (Princeton: Princeton University Press, 1967).

60. Ibid., 182.

61. Ibid., 5.

62. Ernst Mayr, "Avifauna: Turnover on Islands," *Science* 150 (1965): 1587–88. He cites his *Systematics and the Origin of Species* (New York: Columbia University Press, 1942) for the original statement of the idea.

63. Ernst Mayr, *Evolution and the Diversity of Life: Selected Essays* (Cambridge, Mass., and London: Belknap Press, 1976), 616.

64. Frank W. Preston, "The Canonical Distribution of Commonness and Rarity," *Ecology* 43 (1962): 185–215; 410–32.

65. Cole, "Population Consequences of Life History Phenomena," 103–37; Richard Lewontin, "Selection for Colonizing Ability," in *The Genetics of Colonizing Species*, ed. H. G. Baker and G. L. Stebbins (New York: Academic Press, 1965), 79–94.

66. MacArthur and Wilson, *Island Biogeography*, 95–97. The meaning of coarse- and fine-grained used here was slightly different from the original usage by MacArthur and Levins. The concept of grain is also related to Hutchinson's idea of a community which is "homogeneously diverse," as discussed in the context of the broken-stick model.

67. MacArthur, "Some Generalized Theorems of Natural Selection."

68. Mayr, "Avifauna: Turnover on Islands."

69. David Lack, *Island Biology, Illustrated by the Land Birds of Jamaica* (Berkeley and Los Angeles: University of California Press, 1976), 7.

70. Ibid., 6.

71. MacArthur, *Geographical Ecology*, 239.

72. Thomas Schoener, review of *Geographical Ecology*, by R. H. MacArthur, *Science* 178 (1972), 389. Other reviews are: Scott A. Boorman, *Science* 178 (1972): 391–94; Harold Heatwole, *American Scientist* 56 (1968): 150A; François Vuilleumier, *Quarterly Review of Biology* 49 (1974): 163–64; Mark Williamson, *Journal of Animal Ecology* 43 (1974): 601–2. Reviews of *Island Biogeography* are: Frank Preston, *Ecology* 49 (1968): 592–94; Terrell Hamilton, *Science* 159 (1968): 71–72; Mark Williamson, *Journal of Animal Ecology* 38 (1969): 464; D. R. Stoddart, *Nature* 221 (1969): 781–82. See also G. Kolata, "Theoretical Ecology: Beginnings of a Predictive Science," *Science* 183 (1974): 400–1; 450.

73. G. E. Hutchinson, "Variations on a Theme by Robert MacArthur," in *Ecology and Evolution of Communities*, ed. Cody and Diamond, 516.

74. Stephen D. Fretwell, "The Impact of Robert MacArthur on Ecology," *Annual Review of Ecology and Systematics* 6 (1975): 1–13.

75. George W. Salt, "Roles: Their Limits and Responsibilities in Ecological and Evolutionary Research," *American Naturalist* 122 (1983), 698.

76. Eric R. Pianka, *Evolutionary Ecology*, 2d ed. (New York: Harper and Row, 1978); Robert E. Ricklefs, *Ecology* (Newton, Mass.: Chiron, 1973); idem, *The Economy of Nature: A Textbook in Basic Ecology* (Portland, Or.: Chiron, 1976); John Maynard Smith, *Models in Ecology* (Cambridge: Cambridge University Press, 1974).

77. A good idea of the scope of theoretical ecology and the extent to which MacArthur's ideas have been used and refined may be obtained from *Ecology and Evolution of Communities*, ed. Cody and Diamond. See also Robert M. May, ed., *Theoretical Ecology, Principles and Applications*, 2d ed. (Sunderland, Mass.: Sinauer Associates, 1981); idem, *Stability and Complexity in Model Ecosystems*, 2d ed. (Princeton: Princeton University Press, 1974).

78. E. O. Wilson and E. O. Willis, "Applied Biogeography," in *Ecology and Evolution of Communities*, ed. Cody and Diamond, 522–34. For criticisms, see D. Simberloff and L. G. Abele, "Island Biogeography Theory and Conservation Practice," *Science* 191 (1976): 285–86. See the response to their criticism, with their reply, in the series of notes "Island Biogeography and Conservation: Strategy and Limitations," *Science* 193 (1976): 1027–32.

79. R. Levins, review of *Ecology and Resource Management, A Quantitative Approach*, by Kenneth E. F. Watt, *Quarterly Review of Biology* 43 (1968): 301–5.

80. Robert H. MacArthur and Joseph Connell, *The Biology of Populations* (New York: Wiley, 1966). See favorable reviews by V. C. Wynne-Edwards, *American Scientist* 55 (1967): 356A–357A; Leigh Van Valen, *Quarterly Review of Biology* 42 (1967): 524–25; and the highly critical review by L. B. Slobodkin, *Science* 154 (1966): 999.

81. Robert H. MacArthur, "Coexistence of Species," in *Challenging Biological Problems*, ed. J Behnke (New York: Oxford University Press, 1972), 253.

82. MacArthur, "Patterns of Terrestrial Bird Communities," 219. See also W. R. Albury, "The Politics of Truth: A Social Interpretation of Scientific Knowledge, With an Application to the Case of Sociobiology," in *Nature Animated*, ed. Michael Ruse (Dordrecht: D. Reidel, 1982), 115–29. Using the example of sociobiology, Albury argues that the competitive struggle for power and influence within a scientific community is the "principal determining factor in the acceptance of scientific ideas and their certification as 'true' scientific knowledge."

83. MacArthur, "Coexistence of Species," 256.

84. Richard C. Lewontin, ed., *Population Biology and Evolution* (see n.47, above), 2.

85. William R. Thompson, "Notes on the Volterra Equations," *Canadian Entomologist* 92 (1960): 582–94.

86. Kenneth E. F. Watt, "Use of Mathematics in Population Ecology," *Annual Review of Entomology* 7 (1962): 243–60.

87. Kenneth E. F. Watt, ed., *Systems Analysis in Ecology* (New York: Academic Press, 1966), 253–67.

88. Watt, "Use of Mathematics in Population Ecology." Compare his less optimistic assessment ten years later in K. E. F. Watt, "Critique and Comparison of Biome Ecosystem Modeling," in *Systems Analysis and Simulation Ecology*, ed. Bernard C. Patten (New York: Academic Press, 1975), vol. 3, 139–52.

89. Lawrence B. Slobodkin, "Toward a Predictive Theory of Evolution," in *Population Biology and Evolution*, ed. Lewontin, 187–215.

90. Warren Weaver, "The Mathematics of Communication," *Scientific American* 181 (1949): 11–15.

91. Robert H. MacArthur, "Fluctuations of Animal Populations and a Measure of Community Stability," *Ecology* 36 (1955): 533–36. He drew an analogy between information, his stability function for a community, and the functions measuring entropy in Maxwell-Boltzmann statistics, which others later interpreted as having more significance than it deserved. See Henry Horn, "The Ecology of Secondary Succession," *Annual Review of Ecology and Systematics* 4 (1974): 25–37. In 1948 Hutchinson took part in a conference on feedback mechanisms in which Norbert Wiener spoke on information theory. MacArthur would have been familiar with these papers, published in *Annals of the New York Academy of Sciences* 50 (1948). Uses of information theory in ecosystem analysis were discussed by B. C. Patten, "An Introduction to the Cybernetics of the Ecosystem: The Trophic Dynamic Aspect," *Ecology* 40 (1959): 221–31; Ramon Margalef, "Information Theory in Ecology," *General Systems* 3 (1958): 36–71.

92. Pianka, *Evolutionary Ecology*, xii.

93. Ibid.

94. MacArthur, "Coexistence of Species," 257.

95. Pianka, *Evolutionary Ecology*, 267–68.

96. W. D. Hamilton, review of *Evolution in Changing Environments*, by Richard Levins, *Science* 167 (1970): 1478–80.

97. G. W. Salt, "Roles: Their Limits and Responsibilities," 699. A survey of opinions about the value of theory in ecology is in Robert P. McIntosh, "Ecology Since 1900," in *Issues and Ideas in America*, ed. Benjamin J. Taylor and Thurman J. White (Normal, Ok.: University of Oklahoma Press, 1976), 353–72. See also Lawrence B. Slobodkin, "On the Present Incompleteness of Mathematical Ecology," *American Scientist* 53 (1965): 347–57.

98. Leigh Van Valen and Frank A. Pitelka, "Commentary—Intellectual Censorship in Ecology," *Ecology* 55 (1974): 925–26.

99. May, "On Relationships among Various Types of Population Models," 46–57; idem, "The Role of Theory in Ecology," *American Zoologist* 21 (1981): 903–10; E. C. Pielou, "The Usefulness of Ecological Models: A Stock-Taking," *Quarterly Review of Biology* 56 (1981): 17–31.

100. MacArthur, "Coexistence of Species," 259.

101. May, *Stability and Complexity in Model Ecosystems*, 37–38.

102. N. G. Hairston, "Species Abundance and Community Organization," *Ecology* 40 (1959): 404–16.

103. Nelson G. Hairston, "Studies on the Organization of Animal Communities," *Journal of Ecology* (Suppl.) 52 (1964): 227–39.

104. Nelson G. Hairston, "On the Relative Abundance of Species," *Ecology* 50 (1969): 1091–94.

105. Robert M. May, "Patterns of Species Abundance and Diversity," in *Ecology and Evolution of Communities*, eds. Cody and Diamond, 81–120.

106. Richard S. Miller, "Pattern and Process in Competition," *Advances in Ecological Research* 4 (1967): 1–74.

107. Robert H. MacArthur, "Strong, or Weak, Interactions?" *Transactions of the Connecticut Academy of Arts and Sciences* 44 (1972): 179–88.

108. Thomas W. Schoener, "The Controversy over Interspecific Competition," *American Scientist,"* 70 (1982): 586–95.

109. Roger Lewin, "Santa Rosalia Was a Goat," *Science* 221 (1983): 636–39; Paul Harvey and Jonathan Silvertown, "Can Theoretical Ecology Keep a Competitive Edge?" *New Scientist* (15 September 1983): 160–63; Jared Diamond, "Niche Shifts and the Rediscovery of Interspecific Competition," *American Scientist* 66 (1978): 322–31. For the scientific arguments on both sides, see the articles in *American Naturalist* 122 (1983): 583–705.

110. V. Louise Roth, "Constancy in the Size Ratios of Sympatric Species," *American Naturalist* 118 (1981): 394–404.

111. Nelson G. Hairston, "The Experimental Test of an Analysis of Field Distribution: Competition in Terrestrial Salamanders," *Ecology* 61 (1980): 817–26. See also Nelson G. Hairston, "An Experimental Test of a Guild: Salamander Competition," *Ecology* 62 (1981): 65–72.

112. A highly positive review of the theory's impact is in Daniel S. Simberloff, "Equilibrium Theory of Island Biogeography and Ecology," *Annual Review of Ecology and Systematics* 5 (1974): 161–82. For a later critical view see Daniel S. Simberloff, "Using Island Biogeographic Distributions to Determine If Colonization Is Stochastic," *American Naturalist* 112 (1978): 713–26; see also F. S. Gilbert, "The Equilibrium Theory of Island Biogeography: Fact or Fiction?" *Journal of Biogeography* 7 (1980): 209–35; and several of the essays in Alan H. Brush and George A. Clark, Jr., eds., *Perspectives in Ornithology* (Cambridge: Cambridge University Press, 1983).

113. Paul H. Harvey, R. K. Colwell, J. W. Silvertown, and R. M. May, "Null Models in Ecology," *Annual Review of Ecology and Systematics* 14 (1983): 189–211.

114. Daniel Simberloff, "A Succession of Paradigms in Ecology: Essentialism to Materialism to Probabilism," *Synthese* 43 (1980): 3–39.

Notes to Conclusion

1. Richard Levins, *Evolution in Changing Environments: Some Theoretical Explorations* (Princeton: Princeton University Press, 1968); Henry S. Horn, *Adaptive Geometry of Trees* (Princeton: Princeton University Press, 1971); Robert M. May, *Stability and Complexity in Model Ecosystems* (Princeton: Princeton University Press, 1973); Michael E. Gilpin, *Group Selection in Predator-Prey Communities* (Princeton: Princeton University Press, 1975); Joel E. Cohen, *Food Webs and Niche Space* (Princeton: Princeton University Press, 1978).

2. Bernard C. Patten, ed., *Systems Analysis and Simulation in Ecology*, multivolume series (New York: Academic Press), volume 1 in 1971 and others continuing thereafter.

3. John L. Harper, J. N. Clatworthy, I. H. McNaughton, and G. R. Sagar, "The Evolution and Ecology of Closely Related Species Living in the Same Area," *Evolution* 15 (1961): 209–27; J. L. Harper, "The Concept of Population in Modular Organisms," in *Theoretical Ecology*, 2d ed., ed. May, 53–77.

4. Edward O. Wilson, *Sociobiology: The New Synthesis* (Cambridge, Mass.: Harvard University Press, 1975).

5. R. M. Anderson, "Population Ecology of Infectious Disease Agents," in

Theoretical Ecology, 2d ed., ed. May 318–55, and citations therein. R. M. Anderson and R. M. May, eds., *Population Biology of Infectious Diseases* (New York: Springer, 1982); R. M. Anderson, ed., *The Population Dynamics of Infectious Diseases* (London: Chapman and Hall, 1982).

6. May, ed., *Theoretical Ecology*, 17.

7. Louis B. Rosenblatt, "Fossils and Myths: A Comparative Study of Geology and Classical History in Early Victorian England" (Ph.D. diss., Johns Hopkins University, 1983).

8. Charles Elton, review of a symposium on "Plant and Animal Communities," *Journal of Animal Ecology* 9 (1940), 151–52.

9. Ernst Mayr, *Evolution and the Diversity of Life, Selected Essays* (Cambridge, Mass. and London: Belknap Press, 1976). See the essays in Part IV, "Philosophy of Biology."

10. George F. Oster and Edward O. Wilson, *Caste and Ecology in the Social Insects* (Princeton: Princeton University Press, 1978), 295.

11. Ibid., 311.

Select Bibliography

In this bibliography I have listed the main primary published sources consulted with no attempt at completeness. For the modern period the distinction between primary and secondary becomes blurred. Autobiographies are listed among primary sources, but biographical notices or obituaries are among the secondary sources. Secondary sources also include works in the history of ecology, textbooks in ecology, recent review articles in mathematical ecology, and works of general history in this period.

Manuscript Sources

Alfred James Lotka Papers. Princeton University Archives, Princeton, New Jersey (Lotka Papers).

Raymond Pearl Papers (1895–1940). American Philosophical Society Archives, Philadelphia, Pennsylvania (Pearl Papers).

Rockefeller Foundation Archives. Files on the Institute for Biological Research in Boxes 145–146, Record Group 1.1, Series 200D, Johns Hopkins University—Biological Research, 1924–. Rockefeller Archive Center, North Tarrytown, New York (RF Archives).

University of Minnesota, Division of Entomology and Economic Zoology Papers, 1890–1959. University of Minnesota Archives, Minneapolis, Minnesota.

Published Primary Sources

Adams, Charles C. 1913. *Guide to the Study of Animal Ecology*. New York: Macmillan.

———. 1918. "Migration As a Factor in Evolution: Its Ecological Dynamics." *American Naturalist* 52:465–90.

———. 1935. "The Relation of General Ecology to Human Ecology." *Ecology* 16:316–35.

Allee, Warder Clyde. 1931. *Animal Aggregations: A Study in General Sociology*. Chicago: University of Chicago Press.

Allee, W. C.; Emerson, A. E.; Park, O.; Park, T.; and Schmidt, K. P. 1949. *Principles of Animal Ecology*. Philadelphia and London: W. B. Saunders.

Andrewartha, H. G., and Birch, L. C. 1954. *The Distribution and Abundance of Animals*. Chicago and London: University of Chicago Press.

Baas Becking, L. G. M. 1948. "On the Analysis of Sigmoid Curves." *Acta Biotheoretica* 8:42–59.

Bertalanffy, Ludwig von. 1973. *General System Theory: Foundations, Development, Applications*. Rev. ed. New York: George Braziller.

Birch, L. Charles. 1948. "The Intrinsic Rate of Natural Increase of an Insect Population." *Journal of Animal Ecology* 17:15–26.

Bodenheimer, Friedrich Simon. 1938. *Problems of Animal Ecology*. Oxford: Oxford University Press.

———. 1958. *Animal Ecology Today*, vol. 6, *Monographiae Biologicae*. The Hague: W. Junk.

Carr-Saunders, Alex Morris. 1922. *The Population Problem: A Study in Human Evolution*. Oxford: Clarendon.

Chapman, Royal N. 1928. "The Quantitative Analysis of Environmental Factors." *Ecology* 9:111–22.

———. 1931. *Animal Ecology, With Especial Reference to Insects*. New York and London: McGraw-Hill.

———. 1939. "Insect Population Problems in Relation to Insect Outbreak." *Ecological Monographs* 9:261–69.

Chitty, Denis. 1957. "Population Studies and Scientific Methodology." *British Journal for the Philosophy of Science* 8:64–66.

Clements, Frederic E. 1916. *Plant Succession, An Analysis of the Development of Vegetation*. Publication no. 242. Washington D.C.: Carnegie Institute.

Clements, F. E.; Weaver, J. E.; and Hanson, H. C. 1929. *Plant Competition: An Analysis of Community Functions*. Washington, D.C.: Carnegie Institute.

Cold Spring Harbor Symposia on Quantitative Biology. 1957. "Population Studies: Animal Ecology and Demography." Vol. 22.

Cole, Lamont C. 1951. "Population Cycles and Random Oscillations." *Journal of Wildlife Management* 15:233–52.

———. 1954. "The Population Consequences of Life-History Phenomena." *Quarterly Review of Biology* 29:103–37.

Crombie, Alistair C. 1947. "Interspecific Competition." *Journal of Animal Ecology* 16:44–73.

Darwin, Charles Robert. 1964. *On the Origin of Species. A Facsimile of the First Edition*. Cambridge, Mass., and London: Harvard University Press.

de Beer, Gavin R., ed. 1938. *Evolution: Essays on Aspects of Evolutionary Biology Presented to Professor E. S. Goodrich*. Oxford: Clarendon.

Deevey, Edward S. 1947. "Life Tables for Natural Populations of Animals." *Quarterly Review of Biology* 22:283–314.

East, Edward M. 1923. *Mankind at the Crossroads*. New York: Charles Scribners' Sons.

Elton, Charles S. 1924–25. "Periodic Fluctuations in the Numbers of Animals: Their Causes and Effects." *Journal of Experimental Biology* (British) 2: 119–63.

———. 1927. *Animal Ecology*. London: Sidgwick and Jackson. 2d ed. with additional notes; New York: Macmillan, 1935.

———. 1946. "Competition and the Structure of Ecological Communities." *Journal of Animal Ecology* 15:54–68.

Feller, Willy. 1940. "On the Logistic Law of Growth and Its Empirical Verifications in Biology." *Acta Biotheoretica* 5:51–65.

Fisher, Ronald A. 1930. *The Genetical Theory of Natural Selection*. Oxford: Oxford University Press. 2d ed., rev., New York: Dover, 1958.

Forbes, Stephen A. 1907. "History of the Former State Natural History Societies of Illinois." *Science* 26:895.
———. 1977. *Ecological Investigations of Stephen Alfred Forbes*. New York: Arno.
Gause, Georgii Frantsevich. 1934. *The Struggle for Existence*. Baltimore: Williams and Wilkins. Reprint, New York: Dover, 1971.
———. 1935. *Vérifications expérimentales de la théorie mathématique de la lutte pour la vie*. Paris: Hermann.
———. 1936. "The Principles of Biocoenology." *Quarterly Review of Biology* 11:320–36.
———. 1937. "Experimental Populations of Microscopic Organisms." *Ecology* 18:173–79.
———. 1947. "Problems of Evolution." *Transactions of the Connecticut Academy of Arts and Sciences* 37:17–68.
Gause, G. G., and Witt, A. A. 1935. "Behavior of Mixed Populations and the Problem of Natural Selection." *American Naturalist* 69:596–609.
Gray, James. 1928–29. "The Kinetics of Growth." *Journal of Experimental Biology* (British) 6:248–74.
Hairston, Nelson G. 1959. "Species Abundance and Community Organization." *Ecology* 40:404–16.
Haldane, J. B. S. 1924. "A Mathematical Theory of Natural and Artificial Selection." *Transactions of the Cambridge Philosophical Society* 23:19–41.
———. 1932. *The Causes of Evolution*. New York and London: Harper & Brothers.
Hardin, Garrett. 1960. "The Competitive Exclusion Principle." *Science* 131:1292–97.
Harper, John L. 1961. "Approaches to the Study of Plant Competition." *Mechanisms in Biological Competition*. Symposium in Experimental Biology, no. 15. Cambridge: Cambridge University Press.
Hogben, Lancelot. 1931. "Some Biological Aspects of the Population Problem." *Biological Reviews* 6:163–80.
Howard, Leland Ossian. 1919. "Entomology and the War." *Scientific Monthly* 8:109–17.
Howard, L. O., and Fiske, William. "The Importation into the United States of the Parasites of the Gipsy Moth and the Browntail Moth." Bulletin no. 91. Washington, D.C.: U. S. Department of Agriculture, Bureau of Entomology.
Hutchinson, George Evelyn. 1942. "Nati sunt mures, et facta est confusio." Review of Charles Elton's *Voles, Mice, and Lemmings*. *Quarterly Review of Biology* 17:354–57.
———. 1948. "Circular Causal Systems in Ecology." *Annals of the New York Academy of Sciences* 50: 221–46.
———. 1953. "The Concept of Pattern in Ecology." *Proceedings of the Academy of Natural Sciences of Philadelphia* 105:1–12.
———. 1957. "Concluding Remarks." *Cold Spring Harbor Symposia on Quantitative Biology* 22:415–27.
———. 1959. "Homage to Santa Rosalia, or Why Are There So Many Kinds of Animals?" *American Naturalist* 93:145–59.

———. 1961. "The Paradox of the Plankton." *American Naturalist* 95:137–45.
———. 1965. *The Ecological Theater and the Evolutionary Play*. New Haven and London: Yale University Press.
———. 1979. *The Kindly Fruits of the Earth: Recollections of an Embryo Ecologist*. New Haven and London: Yale University Press.
Hutchinson, G. E., and Deevey, E. S., Jr. 1949. "Ecological Studies on Populations." *Survey of Biological Progress* (New York: Academic Press), vol. 1, 325–59.
Huxley, Julian Sorell. 1942. *Evolution: The Modern Synthesis*. London: George Allen and Unwin. American ed.; New York: Harper, 1943.
Jeffreys, Harold. 1931. *Scientific Inference*. Cambridge: Cambridge University Press.
Kermack, W. O., and McKendrick, A. G. 1927. "A Contribution to the Mathematical Theory of Epidemics." *Proceedings of the Royal Society of London*, ser. A, 115:700–721.
Knibbs, George H. 1917. "A Mathematical Theory of Populations, of Its Character and Fluctuations, and of the Factors Which Influence Them." *Census of the Commonwealth of Australia for 1911*, vol. 1, appendix A.
———. 1926. "The Laws of Growth of a Population." *Journal of the American Statistical Association* 21:381–98; 22 (1927):49–59.
Kostitzin, Vladimir A. 1934. *Symbiose, parasitisme et évolution (étude mathématique)*. Paris: Hermann.
———. 1937. *Biologie mathématique*. Paris: Armand Colin.
Lack, David L. 1947. *Darwin's Finches: An Essay on the General Biological Theory of Evolution*. Cambridge: Cambridge University Press.
———. 1954. *The Natural Regulation of Animal Numbers*. Oxford: Clarendon.
———. 1966. *Population Studies of Birds*. Oxford: Clarendon.
———. 1973. "My Life As an Amateur Ornithologist." *Ibis* 115:421–31.
———. 1976. *Island Biology, Illustrated by the Land Birds of Jamaica*. Berkeley and Los Angeles: University of California Press.
Leslie, Patrick H. 1945*a*. "On the Use of Matrices in Certain Population Mathematics." *Biometrika* 33:183–212.
———. 1945*b*. "Some Further Notes on the Use of Matrices in Population Mathematics." *Biometrika* 35:213–45.
Lotka, Alfred James. 1907. "Studies on the Mode of Growth of Material Aggregates." *American Journal of Science*, 4th ser., 24:199–216.
———. 1912. "Evolution in Discontinuous Systems." *Journal of the Washington Academy of Sciences* 2:2–6; 49–59; 66–74.
———. 1914. "An Objective Standard of Value Derived from the Principle of Evolution." *Journal of the Washington Academy of Sciences* 4:409–18; 447–57; 499–500.
———. 1915. "Efficiency As a Factor in Organic Evolution." *Journal of the Washington Academy of Sciences* 5:360–68; 397–403.
———. 1922*a*. "Contribution to the Energetics of Evolution." *Proceedings of the National Academy of Sciences of the U. S. A.* 8:147–51.
———. 1922*b*. "Natural Selection As a Physical Principle." *Proceedings of the National Academy of Sciences of the U. S. A.* 8:151–54.

———. 1923. "Contributions to the Analysis of Malaria Epidemiology." *American Journal of Hygiene* (Suppl.) 3:1–121. In 5 parts; part 4 with F. R. Sharpe.
———. 1925. *Elements of Physical Biology*. Baltimore: Williams and Wilkins. Reprinted with corrections and bibliography as *Elements of Mathematical Biology*. New York: Dover, 1956.
———. 1932*a*. "The Growth of Mixed Populations: Two Species Competing for a Common Food Supply." *Journal of the Washington Academy of Sciences* 22:461–69.
———. 1932*b*. "Contribution to the Mathematical Theory of Capture. I. Conditions for Capture." *Proceedings of the National Academy of Sciences of the U. S. A.* 18:172–78.
———. 1932*c*. "The Structure of a Growing Population." In *Problems of Population*, ed. G. H. L. F. Pitt-Rivers. London: George Allen and Unwin.
———. 1934–39. *Théorie analytique des associations biologiques*. 2 vols. Paris: Hermann.
———. 1939. "Contact Points of Population Study with Related Branches of Science." *Proceedings of the American Philosophical Society* 80:601–26.
———. 1943. "The Place of the Intrinsic Rate of Natural Increase in Population Analysis." *Proceedings of the Eighth American Scientific Congress* 8:297–313.
———. 1945*a*. "Population Analysis As a Chapter in the Mathematical Theory of Evolution." In *Essays on Growth and Form Presented to D'Arcy Wentworth Thompson*, ed. W. E. LeGros Clark and P. B. Medawar. Oxford: Clarendon.
———. 1945*b*. "The Law of Evolution As a Maximal Principle." *Human Biology* 17:167–94.
MacArthur, Robert H. 1957. "On the Relative Abundance of Bird Species." *Proceedings of the National Academy of Sciences of the U. S. A.* 43:293–95.
———. 1958. "Population Ecology of Some Warblers of Northeastern Coniferous Forests." *Ecology* 39:599–619.
———. 1960. "On the Relative Abundance of Species." *American Naturalist* 94:25–36.
———. 1965. "Patterns of Species Diversity." *Biological Reviews* 40:510–33.
———. 1966. "The Theory of the Niche." In *Population Biology and Evolution*, ed. Richard C. Lewontin. Syracuse, N.Y.: Syracuse University Press.
———. 1972*a*. "Coexistence of Species." In *Challenging Biological Problems*, ed. J. Behnke. New York: Oxford University Press.
———. 1972*b*. *Geographical Ecology: Patterns in the Distribution of Species*. New York: Harper & Row.
———. 1972*c*. "Strong, or Weak, Interactions?" *Transactions of the Connecticut Academy of Arts and Sciences* 44:179–88.
MacArthur, R. H., and Levins, Richard. 1964. "Competition, Habitat Selection, and Character Displacement in a Patchy Environment." *Proceedings of the National Academy of Sciences of the U. S. A.* 51:1207–10.
———. 1967. "The Limiting Similarity, Convergence and Divergence of Coexisting Species." *American Naturalist* 101:377–85.
MacArthur, R. H., and Wilson, Edward O. 1967. *The Theory of Island Biogeography*. Princeton: Princeton University Press.
MacLagan, D. Stewart. 1941. "Recent Animal-Population Studies; And Their

Significance in Relation to Socio-Biological Philosophy (Part I)." *Proceedings of the Durham University Philosophical Society* 10:312–31.

Margalef, Ramon. 1958. "Information Theory in Ecology." *General Systems* 3: 36–71.

Mayr, Ernst. 1959. "Where Are We?" *Genetics and Twentieth Century Darwinism. Cold Spring Harbor Symposia on Quantitative Biology* 24:1–14.

Neyman, Jerzy; Park, T.; and Scott, E. L. 1956. "Struggle for Existence. The *Tribolium* Model: Biological and Statistical Aspects." *Proceedings of Third Berkeley Symposium on Mathematical Statistics and Probability* 4:41–79.

Nicholson, Alexander John. 1933. "The Balance of Animal Populations." *Journal of Animal Ecology* 2:132–78.

———. 1954*a*. "Compensatory Reactions of Populations to Stresses, and Their Evolutionary Significance." *Australian Journal of Zoology* 2:1–8.

———. 1954*b*. "An Outline of the Dynamics of Animal Populations." *Australian Journal of Zoology* 2:9–65.

Nicholson, A. J., and Bailey, V. A. 1935. "The Balance of Animal Populations." *Proceedings of the Zoological Society of London,* part 3, pp. 551–98.

Ostwald, Wilhelm. 1902. *Vorlesungen über Naturphilosophie*. Leipzig.

Park, Thomas. 1937. "Experimental Studies of Insect Populations." *American Naturalist* 71:21–33.

———. 1939. "Analytical Population Studies in Relation to General Ecology." *American Midland Naturalist* 21:235–53.

———. 1942. "Integration in Infra-social Insect Populations." *Biological Symposia* 8:121–38.

———. 1946. "Some Observations on the History and Scope of Population Ecology." *Ecological Monographs* 16:315–20.

Park, T.; Leslie, P. H.; and Mertz, D. B. 1964. "Genetic Strains and Competition in Populations of *Tribolium*." *Physiological Zoology* 37:97–162.

Pearl, Raymond. 1909. "Some Recent Studies on Growth." *American Naturalist* 43:302–16.

———. 1915. *Modes of Research in Genetics*. New York: Macmillan.

———. 1922*a*. *The Biology of Death*. Philadelphia: Lippincott.

———. 1922*b*. "The Trends of Modern Biology." *Science* 56:581–92.

———. 1923. "The Interrelations of the Biometric and Experimental Methods of Acquiring Knowledge; With Special Reference to the Problem of the Duration of Life." *Metron* 2:697–721.

———. 1924. *Studies in Human Biology*. Baltimore: Williams and Wilkins.

———. 1925. *The Biology of Population Growth*. New York: Alfred A. Knopf.

———. 1927. "The Growth of Populations." *Quarterly Review of Biology* 2: 532–48.

———. 1928. *The Rate of Living; Being an Account of Some Experimental Studies on the Biology of Life Duration*. New York: Alfred A. Knopf.

———. 1932. "The Influence of Density of Population upon Egg Production in *Drosophila melanogaster*." *Journal of Experimental Zoology* 63:57–84.

———. 1939. *The Natural History of Population*. London and New York: Oxford University Press.

Pearl, Raymond, and Parker, Sylvia L. 1922. "On the Influence of Density of

Population upon the Rate of Reproduction in *Drosophila*." *Proceedings of the National Academy of Sciences of the U. S. A.* 8:212–18.

Pearl, Raymond, and Reed, Lowell J. 1920. "On the Rate of Growth of the Population of the United States Since 1790 and Its Mathematical Representation." *Proceedings of the National Academy of Sciences of the U. S. A.* 6: 275–88.

Pearson, Egon S. 1927. "The Application of the Theory of Differential Equations to the Solution of Problems Connected with the Interdependence of Species." *Biometrika* 19:216–22; 242.

Pearson, Karl. 1900. *The Grammar of Science*. 2d ed. London: Adam & Charles Black.

Pérès, Joseph. 1927. "Une application nouvelle des mathématiques à la biologie: La Théorie des associations biologiques, d'après les travaux de M. Vito Volterra." *Revue Générale des Sciences Pures et Appliquées* 38:295–300; 337–41.

Preston, Frank W. 1948. "The Commonness, and Rarity, of Species." *Ecology* 29:254–83.

Richards, Oscar White. 1927–28. "Potentially Unlimited Multiplication of Yeast with Constant Environment, and the Limiting of Growth by Changing Environment." *Journal of General Physiology* 11:525–38.

———. 1928. "The Growth of the Yeast *Saccharomyces cerevisiae*. I. The Growth Curve, Its Mathematical Analysis, and the Effect of Temperature on the Yeast Growth." *Annals of Botany* 42:271–83.

Robertson, F. W., and Sang, J. H. 1944. "The Ecological Determinants of Population Growth in a *Drosophila* Culture." *Proceedings of the Royal Society of London*, ser. B, 132:258–77; 277–91.

Robertson, T. Brailsford. 1923. *The Chemical Basis of Growth and Senescence*. Philadelphia and London: Lippincott.

Robson, G. C., and Richards, O. W. 1936. *The Variation of Animals in Nature*. London: Longmans, Green.

Ross, Ronald. 1911. "Some Quantitative Studies in Epidemiology." *Nature* 87:466–67.

———. 1923. *Memoirs, With a Full Account of the Great Malaria Problem and Its Solution*. London: John Murray.

Salt, George. 1934. "Experimental Studies in Insect Parasitism. I. Introduction and Technique. II. Superparasitism." *Proceedings of the Royal Society of London*, ser. B, 114:450–54; 455–76.

Sanger, Margaret, ed. 1927. *Proceedings of the World Population Conference, Geneva, 1927*. London: Edward Arnold.

Scudo, Francesco M., and Ziegler, James R. 1978. *The Golden Age of Theoretical Ecology, 1923–1940*. Berlin: Springer-Verlag.

Severtzov, S. A. 1934. "On the Dynamics of Populations of Vertebrates." *Quarterly Review of Biology* 9:409–37.

Simberloff, Daniel S. 1978. "Using Island Biogeographic Distributions to Determine If Colonization Is Stochastic." *American Naturalist* 112:713–26.

Simberloff, D. S., and Boecklen, William. 1981. "Santa Rosalia Reconsidered: Size Ratios and Competition." *Evolution* 35:1206–28.

Slobodkin, Lawrence B. 1961. *Growth and Regulation of Animal Populations.* New York: Holt, Rinehart and Winston. 2d ed., rev., New York: Dover, 1980.

Smith, David, and Keyfitz, Nathan, eds. 1977. *Mathematical Demography: Selected Papers.* Berlin: Springer-Verlag.

Smith, Frederick E. 1952. "Experimental Methods in Population Dynamics: A Critique." *Ecology* 33:441–50.

Smith, Harry Scott. 1935. "The Role of Biotic Factors in the Determination of Population Densities." *Journal of Economic Entomology* 28:873–98.

Solomon, M. E. 1949. "The Natural Control of Animal Populations." *Journal of Animal Ecology* 18:1–35.

Spencer, Herbert. 1880. *First Principles.* 4th ed. Rev. and enl. New York: Thomas Y. Crowell.

———. 1896. *The Principles of Biology.* 2 vols. Rev. and enl. ed. New York: D. Appleton.

Stauffer, Robert C., ed. 1975. *Charles Darwin's Natural Selection; Being the Second Part of His Big Species Book Written from 1856 to 1858.* Cambridge: Cambridge University Press.

Stevenson, T. H. C. 1925. "The Laws Governing Population." *Journal of the Royal Statistical Society of London,* ser. A, 88:63–90, incl. discussion.

Thompson, D'Arcy Wentworth. 1917. *On Growth and Form.* Cambridge: Cambridge University Press. 2d ed., Cambridge University Press, 1942.

Thompson, Warren S. 1930. *Population Problems.* New York: McGraw-Hill.

Thompson, William Robin. 1923. "La théorie mathématique de l'action des parasites entomophages." *Revue Générale des Sciences Pures et Appliquées* 34:202–10.

———. 1929. "On Natural Control." *Parasitology* 21.

———. 1930*a*. "The Principles of Biological Control." *Annals of Applied Botany* 17:306–38.

———. 1930*b*. "The Utility of Mathematical Methods in Relation to Work on Biological Control." *Annals of Applied Biology* 17:641–48.

———. 1930*c*. *The Biological Control of Insect and Plant Pests: A Report on the Organisation and Progress of the Work of Farnham House Laboratory.* London: King's Printer.

———. 1937. *Science and Common Sense: An Aristotelian Excursion.* London: Longmans, Green.

———. 1939. "Biological Control and the Theories of the Interactions of Populations." *Parasitology* 31:299–388.

———. 1948. "Can Economic Entomology Be an Exact Science?" *Canadian Entomologist* 80:49–55.

———. 1956. "The Fundamental Theory of Natural and Biological Control." *Annual Review of Entomology* 1:379–402.

Varley, George C. 1947. "The Natural Control of Population Balance in the Knapweed Gall-fly." *Journal of Animal Ecology* 16:139–86.

Verhulst, Pierre-François. 1838. "Notice sur la loi que la population suit dans son accroissement." *Correspondances Mathématiques et Physiques* 10:113–21.

———. 1845. "Recherches mathématiques sur la loi d'accroissement de la popula-

tion." *Nouveaux Mémoires de l'Académie Royale des Sciences et Belles-Lettres de Bruxelles* 18:3–38.
———. 1847. "Deuxième Mémoire sur la loi d'accroissement de la population." *Mémoires de l'Académie Royale des Sciences des Lettres et des Beaux-Arts de Belgique,* ser. 2, 20:3–32.
Vernadsky, Vladimir, I. 1929. *La Biosphère.* Paris: Felix Alcan.
Volterra, Vito. 1926. "Fluctuations in the Abundance of a Species Considered Mathematically." *Nature* 118:558–60.
———. 1928. "Variations and Fluctuations of the Number of Individuals in Animal Species Living Together." *Journal du Conseil International pour l'Exploration de la Mer* 3:3–51.
———. 1931. *Leçons sur la Théorie mathématique de la lutte pour la vie.* Ed. Marcel Brelot. Paris: Gauthier-Villars.
———. 1937. "Principes de biologie mathématique." *Acta Biotheoretica* 3:1–36.
Volterra, Vito, and D'Ancona, Umberto. 1935. *Les associations biologiques au point de vue mathématique.* Paris: Hermann.
Wardle, Robert A. 1929. *The Problems of Applied Entomology.* New York: McGraw-Hill.
Watt, Kenneth E. F. 1962. "Use of Mathematics in Population Ecology." *Annual Review of Entomology* 7:243–60.
———, ed. 1966. *Systems Analysis in Ecology.* New York: Academic Press.
Wilson, E. B. 1926*a*. "Statistical Inference." *Science* 63:290–91.
———. 1926*b*. "Empiricism and Rationalism." *Science* 64:47–57.
Wilson, E. B., and Puffer, Ruth R. 1933. "Least Squares and Laws of Population Growth." *Proceedings of the American Academy of Arts and Sciences* 68:285–382.
Winsor, Charles P. 1934. "Mathematical Analysis of Growth of Mixed Populations." *Cold Spring Harbor Symposia on Quantitative Biology* 2:181–86.
Wolfe, A. B. 1927. "Is There a Biological Law of Human Population Growth?" *Quarterly Journal of Economics* 41:557–94.
Yule, G. Udny. 1925. "The Growth of Population and the Factors Which Control It." *Journal of the Royal Statistical Society of London,* ser. A, 88:1–58.

Secondary Sources

Allen, Garland E. 1978. *Life Science in the Twentieth Century.* Cambridge: Cambridge University Press.
American Plant Ecology, 1897–1917. 1977. New York: Arno.
Bartlett, Maurice S. 1960. *Stochastic Population Models in Ecology and Epidemiology.* London: Methuen.
Bowler, Peter J. 1983. *The Eclipse of Darwinism: Anti-Darwinian Evolution Theories in the Decades around 1900.* Baltimore and London: Johns Hopkins University Press.
Burgess, Robert L. 1981. "Sources of Biographical Information on American Ecologists." *Bulletin of Ecological Society of America* 62:236–55.
Cittadino, Eugene. 1980. "Ecology and the Professionalization of Botany in America, 1890–1905." *Studies in History of Biology* 4:171–98.

Cody, Martin L., and Diamond, Jared M., eds. 1975. *Ecology and Evolution of Communities*. Cambridge, Mass., and London: Belknap Press.

Cox, David L. 1979. *Charles Elton and the Emergence of Modern Ecology*. Ph.D. diss., Washington University.

Cravens, Hamilton. 1978. *The Triumph of Evolution: American Scientists and the Heredity-Environment Controversy, 1900–1941*. Philadelphia: University of Pennsylvania Press.

Currie, G., and Graham, J. 1968. "Growth of Scientific Research in Australia: The Council for Scientific and Industrial Research and the Empire Marketing Board." *Australian Academy of Science Records* 1:25–35.

D'Ancona, Umberto. 1954. *The Struggle for Existence*. Bibliotheca Biotheoretica, vol. 6. Trans. A. Charles and R. F. J. Withers. Leiden: E. J. Brill.

DeBach, Paul. 1966. "The Competitive Displacement and Coexistence Principles." *Annual Review of Entomology* 11:183–212.

———. 1974. *Biological Control by Natural Enemies*. Cambridge: Cambridge University Press.

Diamond, Jared. 1978. "Niche Shifts and the Rediscovery of Interspecific Competition." *American Scientist* 66:322–31.

Doutt, R. L. 1958. "Vice, Virtue and the Vedalia." *Bulletin of the Entomological Society of America* 4:119–23.

Dunlap, Thomas R. 1980. "Farmers, Scientists, and Insects." *Agricultural History* 54:93–107.

———. 1981. *DDT: Scientists, Citizens, and Public Policy*. Princeton: Princeton University Press.

Ecological Studies on Insect Parasitism. 1977. New York: Arno.

Egerton, Frank N. 1973. "Changing Concepts of the Balance of Nature." *Quarterly Review of Biology* 48:322–50.

———. 1977. "A Bibliographical Guide to the History of General Ecology and Population Ecology." *History of Science* 15:189–215.

———. 1983. "The History of Ecology: Achievements and Opportunities, Part One." *Journal of the History of Biology* 16:259–310.

Finerty, James P. 1980. *The Population Ecology of Cycles in Small Mammals: Mathematical Theory and Biological Fact*. New Haven and London: Yale University Press.

Fretwell, Stephen D. 1975. "The Impact of Robert MacArthur on Ecology." *Annual Review of Ecology and Systematics* 6:1–13.

Graham, Samuel A. 1941. "Royal Norton Chapman, 1889–1939." *Annals of the Entomological Society of America* 34:521–24.

Greathead, D. J. 1980. "Biological Control of Pests and the Contributions of the CIBC." *Antenna* 4:88–91.

Greenwood, Major. 1932. *Epidemiology, Historical and Experimental*. Baltimore: Johns Hopkins University Press.

Grene, Marjorie, ed. 1983. *Dimensions of Darwinism: Themes and Counter Themes in Twentieth-Century Evolutionary Theory*. Cambridge: Cambridge University Press.

Hagen, K. S., and Franz, J. M. 1973. "A History of Biological Control." In *History*

of Entomology, ed. R. F. Smith, T. E. Mittler, and C. N. Smith, 433–76. Palo Alto, Calif.: Annual Reviews.
Harper, John L. 1974. "A Centenary in Population Biology." *Nature* 252:526–27.
Harvey, Paul H.; Colwell, R. K.; Silvertown, J. W.; and May, R. M. 1983. "Null Models in Ecology." *Annual Review of Ecological Systematics* 14:189–211.
Harvey, Paul, and Silvertown, Jonathan. 1983. "Can Theoretical Ecology Keep a Competitive Edge?" *New Scientist* 100 (September 15): 160–63.
Hauser, Philip M. 1965. "Demography and Ecology." *Annals of the American Academy of Political and Social Sciences* 362:129–38.
Hauser, Philip M., and Duncan, Otis D. 1959. *The Study of Population: An Inventory and Appraisal*. Chicago and London: University of Chicago Press.
Hays, Samuel P. 1957. *The Response to Industrialism, 1855–1914*. Chicago: University of Chicago Press.
———. 1959. *Conservation and the Gospel of Efficiency: The Progressive Conservation Movement 1890–1920*. Cambridge, Mass.: Harvard University Press.
Hesse, Mary B. 1966. *Models and Analogies in Science*. Notre Dame, Ind.: University of Notre Dame Press.
History of American Ecology. 1977. New York: Arno.
Howard, Leland Ossian. 1931. *A History of Applied Entomology*. Vol. 84, Miscellaneous Collection. Washington, D. C.: Smithsonian Institution.
———. 1932. "Stephen Alfred Forbes, 1844–1930." *Biographical Memoirs of the National Academy of Sciences of the U. S. A.* 15:1–54.
———. 1933. *Fighting the Insects; The Story of an Entomologist*. New York: Macmillan.
Hunsaker, Jerome, and MacLane, Saunders. 1973. "Edwin Bidwell Wilson, 1879–1964." *Biographical Memoirs of the National Academy of Sciences of the U. S. A.* 43:285–320.
Hutchinson, George Evelyn. 1978. *An Introduction to Population Ecology*. New Haven and London: Yale University Press.
———. 1980. "Lorande Loss Woodruff. 1879–1947." *Biographical Memoirs of the National Academy of Sciences of the U. S. A.* 52:471–85.
Jackson, Jeremy B. C. 1981. "Interspecific Competition and Species' Distributions: The Ghosts of Theories and Data Past." *American Zoologist* 21:889–901.
Jennings, Herbert Spencer. 1942. "Raymond Pearl, 1879–1940." *Biographical Memoirs of the National Academy of Sciences of the U. S. A.* 22:295–347.
Joravsky, David. 1970. *The Lysenko Affair*. Cambridge, Mass.: Harvard University Press.
Kingsland, Sharon E. 1982. "The Refractory Model: The Logistic Curve and the History of Population Ecology." *Quarterly Review of Biology* 57:29–52.
Kottler, Malcolm J. 1976. "Isolation and Speciation, 1837–1900." Ph.D. diss., Yale University.
Leigh, Egbert G. 1968. "The Ecological Role of Volterra's Equations." *Some Mathematical Problems in Biology* 1:1–61.
Levins, Richard. 1966. "The Strategy of Model Building in Population Biology." *American Scientist* 54:421–31.

Lewin, Roger. 1983. "Santa Rosalia Was a Goat." *Science* 221:636–39.

Lloyd, P. J. 1967. "American, German, and British Antecedents to Pearl and Reed's Logistic Curve." *Population Studies* 21:99–108.

Lopez, Alvaro. 1961. *Problems in Stable Population Theory*. Princeton, N.J.: Office of Population Research.

Ludmerer, Kenneth M. 1972. *Genetics and American Society; An Historical Appraisal.* Baltimore and London: Johns Hopkins University Press.

MacFadyen, Amyan. 1975. "Some Thoughts on the Behaviour of Ecologists." *Journal of Ecology* 63:379–92.

May, Robert M. 1973. "On Relationships among Various Types of Population Models." *American Naturalist* 107:46–57.

———. 1974. *Stability and Complexity in Model Ecosystems*. 2d ed. Princeton: Princeton University Press.

———, ed. 1976. *Theoretical Ecology, Principles and Applications*. Oxford: Blackwell Scientific Publications. 2d ed; Sunderland, Mass.: Sinauer Associates, 1981.

———. 1981. "The Role of Theory in Ecology." *American Zoologist* 21:903–10.

Mayr, Ernst, and Provine, William B., eds. 1980. *The Evolutionary Synthesis: Perspectives on the Unification of Biology*. Cambridge, Mass., and London: Harvard University Press.

McIntosh, Robert P. 1974. "Plant Ecology, 1947–1972." *Annals of Missouri Botanical Garden* 61:132–65.

———. 1976. "Ecology Since 1900." In *Issues and Ideas in America,* ed. Benjamin J. Taylor and Thurman J. White. Norman, Okla.: University of Oklahoma Press.

———. 1980. "The Background and Some Current Problems of Theoretical Ecology." *Synthese* 43:195–255.

Mertz, David B. 1972. "The *Tribolium* Model and the Mathematics of Population Growth." *Annual Review of Ecology and Systematics* 3:51–78.

Mertz, David B., and McCauley, David E. 1980. "The Domain of Laboratory Ecology." *Synthese* 43: 95–110.

Miner, John Rice, trans. 1933. "Pierre-François Verhulst, The Discoverer of the Logistic Curve." *Human Biology* 5:673–89.

Notestein, Frank W. 1950. "Alfred James Lotka, 1880–1949." *Population Index* 16:22–29.

Odum, Eugene P. 1968. "Energy Flow in Ecosystems: A Historical Review." *American Zoologist* 8:11–18.

———. 1971. *Fundamentals of Ecology,* 3d ed. Philadelphia: W. B. Saunders.

Oleson, Alexandra, and Voss, John. 1979. *The Organization of Knowledge in Modern America, 1860–1920.* Baltimore and London: Johns Hopkins University Press.

Pearsall, W. H. 1964. "The Development of Ecology in Britain." *Journal of Ecology* (Suppl.) 52:1–12.

Pianka, Eric R. 1978. *Evolutionary Ecology*. 2d ed. New York: Harper and Row.

Pielou, E. C. 1969. *An Introduction to Mathematical Ecology*. New York: John Wiley and Sons.

———. 1981. "The Usefulness of Ecological Models: A Stock-Taking." *Quarterly Review of Biology* 56:17–31.

Provine, William B. 1971. *The Origins of Theoretical Population Genetics*. Chicago and London: University of Chicago Press.

Rapport, David J., and Turner, James E. 1977. "Economic Models in Ecology." *Science* 195:367–73.

Russett, Cynthia E. 1966. *The Concept of Equilibrium in American Social Thought*. New Haven and London: Yale University Press.

Salt, George W. 1983. "Roles: Their Limits and Responsibilities in Ecological and Evolutionary Research." *American Naturalist* 122:697–705.

Salvatore, Augusto G. 1972. "On the Applications of Mathematics to Certain Biological Problems Considered by Alfred J. Lotka (1880–1949) and Vito Volterra (1860–1940)." PhD. diss., New York University.

Samuelson, Paul A. 1976. "Resolving a Historical Confusion in Population Analysis." *Human Biology* 48:559–80.

Schoener, Thomas W. 1982. "The Controversy over Interspecific Competition." *American Scientist* 70:586–95.

Scudo, Francesco M. 1971. "Vito Volterra and Theoretical Ecology." *Theoretical Population Biology* 2:1–23.

Scudo, Francesco M., and Ziegler, James R. 1976. "Vladimir Aleksandrovich Kostitzin and Theoretical Ecology." *Theoretical Population Biology* 10:395–412.

———, eds. 1978. *The Golden Age of Theoretical Ecology: 1923–1940. A Collection of Works by Volterra, Kostitzin, Lotka, and Kolmogoroff*. Berlin: Springer-Verlag.

Sears, Paul B. 1956. "Charles C. Adams, Ecologist." *Science* 123:974.

Serfling, Robert E. 1952. "Historical Review of Epidemic Theory." *Human Biology* 24:145–66.

Simberloff, Daniel. 1980. "A Succession of Paradigms in Ecology: Essentialism to Materialism to Probabilism" (with reply). *Synthese* 43:3–39; 79–93.

Simpson, George Gaylord. 1953. "The Baldwin Effect." *Evolution* 7:110–17.

Smith, John Maynard. 1974. *Models in Ecology*. Cambridge: Cambridge University Press.

Stauffer, Robert C. 1957. "Haeckel, Darwin, and Ecology." *Quarterly Review of Biology* 32:138–44.

Stearns, Steven C. 1976. "Life-History Tactics: A Review of the Ideas." *Quarterly Review of Biology* 51:3–47.

Streifer, William. 1974. "Realistic Models in Population Ecology." *Advances in Ecological Research* 8:199–266.

Sulloway, Frank J. 1982. "Darwin and His Finches: The Evolution of a Legend." *Journal of the History of Biology* 15:1–53.

Taylor, L. R., and Elliott, J. M. 1981. "The First Fifty Years of the *Journal of Animal Ecology*—with an Author Index." *Journal of Animal Ecology* 50: 951–71.

Thorpe, William H. 1973. "William Robin Thompson, 1887–1972." *Biographical Memoirs of the Royal Society of London* 19:655–78.

———. 1974. "David Lambert Lack, 1910–1973." *Biographical Memoirs of the Royal Society of London* 20:271–93.

Turner, Thomas B. 1974. *Heritage of Excellence: The Johns Hopkins Medical Institutions 1914–1947*. Baltimore and London: Johns Hopkins University Press.

Vandermeer, John H. 1972. "Niche Theory." *Annual Review of Ecology and Systematics* 3:107–32.

Vorzimmer, Peter J. 1965. "Darwin's Ecology and Its Influence upon His Theory." *Isis* 56:148–55.

Wangersky, P. J. 1978. "Lotka-Volterra Population Models." *Annual Review of Ecology and Systematics* 9:189–218.

Weiner, Douglas R. 1983. "The History of the Conservation Movement in Russia and the U.S.S.R. from its Origins to the Stalin Period." Ph.D. diss., Columbia University.

Whittaker, E. T. 1939–41. "Vito Volterra, 1860–1940." *Obituary Notices of the Royal Society of London* 3:691–729.

Wiebe, Robert H. 1967. *The Search for Order, 1877–1920*. New York: Hill and Wang.

Wimsatt, William C. 1980. "Randomness and Perceived-Randomness in Evolutionary Biology." *Synthese* 43:287–329.

———. 1981. "Robustness, Reliability and Multiple Determination in Nature." In *Knowing and Validating in the Social Sciences: A Tribute to Donald T. Campbell,* ed. M. Brewer and B. Collins. San Francisco: Jossey-Bass.

Worster, Donald. 1977. *Nature's Economy: The Roots of Ecology*. San Francisco: Sierra Club.

Worthington, E. Barton, ed. 1975. *The Evolution of IBP*. Cambridge: Cambridge University Press.

———. 1983. *The Ecological Century, A Personal Appraisal*. Oxford: Clarendon.

Index